HIGHLIGHTS IN ENVIRONMENTAL RESEARCH

SERIES ON ENVIRONMENTAL SCIENCE AND MANAGEMENT

Series Editor: Professor J.N.B. Bell
Centre for Environmental Technology, Imperial College

Published

Vol. 1 Environmental Impact of Land Use in Rural Regions
P.E. Rijtema, P. Groenendijk and J.G. Kroes

Vol. 2 Numerical Ocean Circulation Modeling
D.B. Haidvogel and A. Beckmann

Forthcoming

Biology of Wastewater Treatment (New Edn.)
N. F. Gray

The Kraken Sleeps: Chemical Weapons Dumping in the Irish and Baltic Seas
K. R. Seddon and O. A. Smirnova

SERIES ON ENVIRONMENTAL SCIENCE AND MANAGEMENT VOL. 3

HIGHLIGHTS IN ENVIRONMENTAL RESEARCH

Professorial Inaugural Lectures at Imperial College

Editor

Sir John Mason

Centre for Environmental Technology, Imperial College of Science, Technology and Medicine, UK

ICP Imperial College Press

Published by

Imperial College Press
57 Shelton Street
Covent Garden
London WC2H 9HE

Distributed by

World Scientific Publishing Co. Pte. Ltd.
P O Box 128, Farrer Road, Singapore 912805
USA office: Suite 1B, 1060 Main Street, River Edge, NJ 07661
UK office: 57 Shelton Street, Covent Garden, London WC2H 9HE

British Library Cataloguing-in-Publication Data
A catalogue record for this book is available from the British Library.

HIGHLIGHTS IN ENVIRONMENTAL RESEARCH

ISBN 1-86094-103-6

Printed in Singapore.

Preface

The Imperial College of Science, Technology and Medicine conducts a very large programme of research in many areas of environmental science, technology and policy ranging from climate science, through engineering to environmental law. The diversity and flavour of this programme, which involves virtually every department of the College, is illustrated in this collection of inaugural lectures given by newly appointed professors in recent years.

The lectures are presented at a level which I hope will appeal to a broad cross-section of scientists and policy-makers working in this complex and strongly interdisciplinary field.

Editor
John Mason

About the Chapter Contributors

~Professor John Harries~

After obtaining his degree from Birmingham University and his PhD from King's College, London, he worked at National Physical Laboratory as Scientific Officer. He became Deputy Chief Scientific Officer and then Head of the Space Science Department at Rutherford Appleton Laboratory. Over 100 papers and articles have been published, including a book entitled "Earthwatch the climate from space". Currently he is Professor of Earth Observation at Imperial College, London.

~Professor Howard Wheater~

Howard Wheater is Head of Environmental & Water Resource Engineering, Imperial College, and President of the British Hydrological Society. Research interests include surface and groundwater hydrology, water quality and waste management. He has extensive international research and project experience and recently represented Hungary at the International Court of Justice concerning a dispute with Slovakia over Danube dams.

~Professor John Monhemius~

Holds the Roy Wright Chair in Mineral and Environmental Engineering at Imperial College. He has more than 30 years experience of academic research in hydrometallurgy and environmental control in mining and metallurgical processes, particularly in the Medal of the Institution of Mining and Metallurgy for research into the extraction of gold and in 1995, the MIRO John Phillips Medal for innovation and application of technology for the development of stable arsenical compounds for disposal of arsenic. He has acted as a consultant to many large mining and chemical companies.

~Professor Iain Thornton~

Iain Thornton, PhD, DSc, is Professor of Environmental Geochemistry and Director of the Environmental Geochemistry Research Group. He has over thirty years of research experience in environmental geochemistry and geochemical mapping, the chemistry and behaviour of trace elements and metals in soils and waters, and the effects of metal exposure on plant, animal and human health. He is also undertaking research into urban geochemistry in the UK and People's Republic of China. He is the editor of the standard text Applied Environmental Geochemistry, and has published widely on sources and pathways on metals in the environmental and their impacts. He is an elected Member of the Norwegian Academy of Science and Letters and a Fellow of the Geological Society of London. He was President of the Society for Environmental Geochemistry and Health from 1985 to 1987.

~Professor John Woods~

Professor John Woods obtained his PhD in cloud physics at Imperial College in 1964 and then joined the Meteorological Office where his interest gradually moved to oceanography, especially the structure of the ocean surface layers. Soon afterwards he was appointed to chairs of physical oceanography at the Universities of Southampton and Kiel (Germany) and then as Deputy Director of the Natural Environment Research Council in charge of oceanographic and atmospheric sciences. In 1995, he was appointed Dean of the new Graduate School for the Environment at Imperial College where he is currently Professor of Physical Oceanography. It has been deeply involved in the planning of major international oceanographic research programmes over three decades.

~Professor Graham Matthews~

Graduated at Imperial College in 1957 and was a Colonial Office Research Entomologist with the Cotton Pest Research Scheme in the Federation of Rhodesia and Nyasaland until 1967. He returned to IC and since 1972 has been responsible for the International Pesticide Application Research Centre at Silwood Park.

Contents

Preface v

About the Chapter Contributors vii

Chapter 1: THE RAINBOW PLANET: STUDIES OF THE EARTH'S CLIMATE FROM SPACE 1
John E. Harries

Chapter 2: WATER IN A CHANGING WORLD 17
Howard S. Wheater

Chapter 3: MAN, MINERALS AND ENVIRONMENT 67
A.J. Monhemius

Chapter 4: ROCK, SOIL AND HEALTH: MAN'S INTERFERENCE WITH THE GEOCHEMICAL ENVIRONMENT 91
Iain Thornton

Chapter 5: VIRTUAL ECOLOGY 133
John D. Woods

Chapter 6: PESTS, PESTICIDES AND PEST MANAGEMENT 165
Graham A. Matthews

CHAPTER 1

THE RAINBOW PLANET: STUDIES OF THE EARTH'S CLIMATE FROM SPACE*

JOHN E. HARRIES

"I'll put a girdle round about the Earth in forty minutes".
Puck, in *A Midsummer Night's Dream*

"Now there is one outstandingly important fact regarding spaceship Earth, and that is that no instruction book came with it".
R. Buckminster-Fuller, in *Operating Manual for Spaceship Earth*

"Mostly Harmless".
Entry for 'Earth' in *The Hitchhiker's Guide to the Galaxy*

Background

The planet Earth, seen from a distance in space, is predominantly blue and white in colour. However, a closer look will reveal the brown hues of the land; the greens of the vegetation; and the reds and yellows of the land surfaces. Earth Observation uses all these colours to study Planet Earth, plus the invisible wavelengths in the ultra-violet (UV), infra-red

*Inaugural lecture given in the Blackett Laboratory at Imperial College, London, on 12 March 1997.

(IR) and microwaves (MW). A huge range of wavelengths is available if we use this complete spectrum, from the far UV at a wavelength of about 100 nm (10^{-7} m) to the MW at about 1 mm (10^{-3} m). The visible part of the spectrum occupies a really narrow slice of this range — between about 500 nm and 700 nm.

Earth Observation (EO) is primarily carried out from space using instruments to measure radiation from the Earth at these many wavelengths, and using a variety of ingenious techniques that maximise the efficiency of observation from orbiting spacecraft. Measurements are also made, on a more localised basis, from high altitude balloons, aircraft and the ground.

What are the advantages of space? Space provides global, or at least near global observations. Orbiting satellites can circle the Earth in about 100 minutes (so Shakespeare was almost right!) at heights of several hundred kilometres. Geostationary equatorial orbits exist at altitudes of about 36,000 km, at which orbital periods of 24 hours allow a spacecraft to orbit in synchronism with the Earth, thereby appearing to hover over one point on the equator. Other more complex orbits are also possible, which can, to some extent, be tailored to the application in mind (for example, highly elliptical orbits allow a spacecraft to spend a lot of time over one particular part of the Earth's surface as it passes out and back on its journey to apogee).

The instruments on board a spacecraft need to be fast, sensitive and multi-spectral in order to extract all the information available from the radiation passing up from the Earth towards the spacecraft, covering all the colours of the Rainbow Planet. This is because satellites move very quickly (for most polar orbits, around 7 km s^{-1}), and the short time available to observe any one part of the Earth must be used to the full.

The Climate System — A Primer

The climate system comprises the atmosphere, the oceans, the land surface, the cryosphere (regions of snow and ice) and the biosphere (all living things, including vegetation, animals and humans), as well as all interactions between these components. It is, therefore, a fiendishly complex system capable of countless feedbacks between individual processes, many of these feedbacks begin non-linear in nature (see Fig. 1). The system as a whole, of course, is driven by the energy input from the Sun, and since the planet is in thermal equilibrium (as accurately as we can measure), this input of energy is balanced by the output of energy as long-wave IR radiation from the Earth back to the cold background of space. Figure 2 illustrates this balance between the radiation input from the Sun at the Earth's orbit and the output of IR radiation which cools the planet to space.

Despite its complexity, it is both a matter of intellectual challenge and of great practical importance that we attempt to understand the basic scientific laws governing this climate system. Work is going on in many centres around the world. Here, at Imperial College, we are trying to understand the basic physics of the climate system by using observations made from space and elsewhere, as well as advanced computer models of the climate. There are many interacting processes which we have to take into account, as well as vast ranges of time and space scales (seconds to millenia; microns to the planetary diameter of 12,000 km). It is a daunting prospect, but already the scientific community has learned a great about the workings of our planet; in recent years, a lot of this information has come from observations from space.

A central issue is the question of the fragility and sensitivity of the climate system. In fact, the climate seems to be a pretty stable system — from data covering geological time scales — and the changes which concern us, while important to human activity, are rather small when measured as a fractional change in, say, surface temperature, cloudiness or rainfall. However, it is a frightening thought that the atmosphere is effectively about 7 km thick (this is the height at which the density has fallen to 1/e of that at the surface). This thickness corresponds to a layer of about 0.1% of the radius of the Earth. Thus, the atmosphere which blankets and protects the surface from harsh UV radiation, and

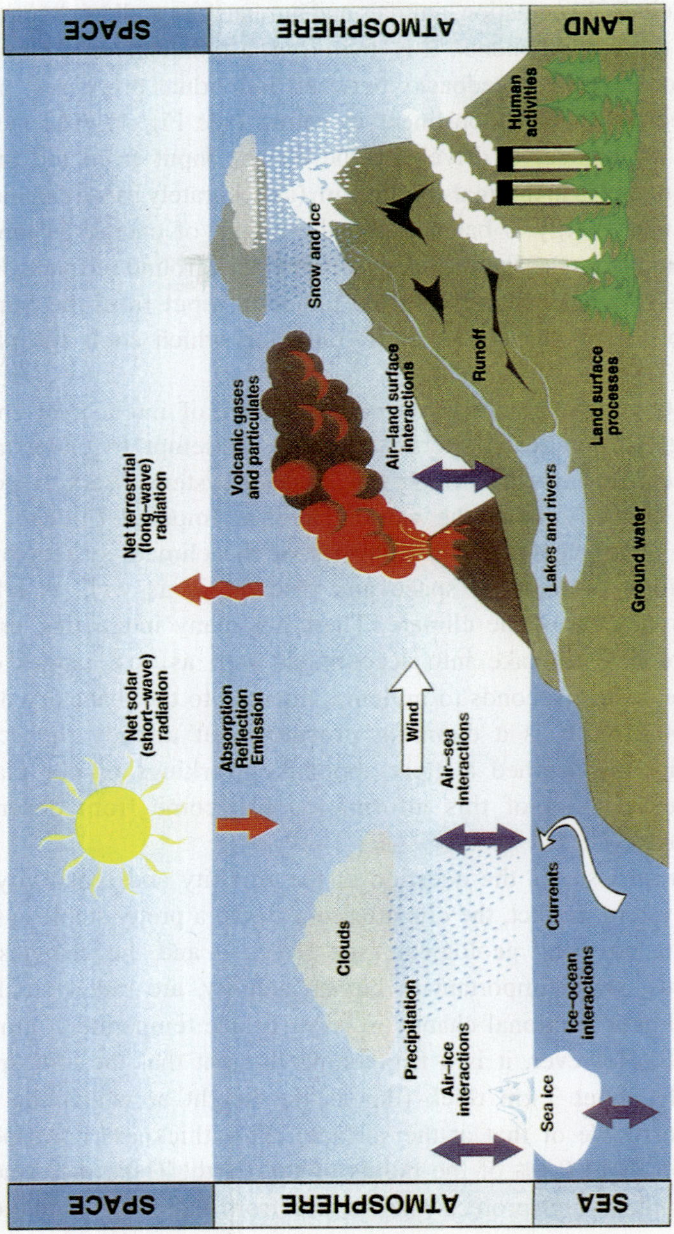

Fig. 1. Schematic diagram of the interactions within the climate system.

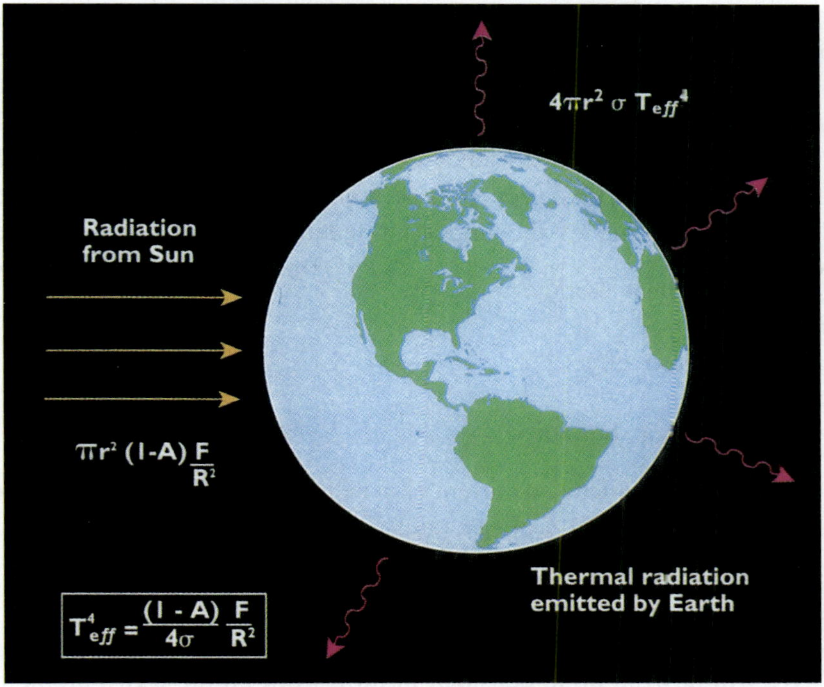

Fig. 2. The radiative balance of a thermally black Earth: the effective equilibrium radiative temperature, T_{eff}, is defined by equating the incoming energy per second $[\pi r^2(1 - A)F/R^2]$ to the thermal energy loss per second $[4\pi r^2\sigma T^4_{eff}]$. r is the Earth radius; A is the planetary albedo; F is the solar constant (1370 W m^{-2}), R is the sun-earth distance (= 1 AU); and σ is the Stefan-Boltzmann constant.

from extremes of temperature during the diurnal cycle, is an extremely thin shell that would appear to be quite vulnerable to change. We do not really understand how vulnerable.

The Earth is uniquely the "Water Planet". Water plays a profoundly important role in many aspects of the operation of the planet. The existence of significant quantities of water in all three phases (solid, liquid and vapour) is of critical importance to the Earth's geophysical and biospheric systems. A great part of the physical and biological diversity we observe and enjoy on our planet is due to the presence of water: it gives rise to clouds and rain, plant growth, animal development, oceanic processes and biota, and the variety of landscape and ecosystems. Within the total Earth system, the balance — and indeed the long-term stability of the climate — is due to the existence of a massive thermal "buffer" in the oceans and the radiative properties of water vapour, ice and liquid water clouds. Thus, some of the important features for our topic include:

(i) Oceanic heat storages;
(ii) Ice sheets;
(iii) Greenhouse effect;
(iv) Clouds;
(v) Atmospheric humidity.

Two major problems in current climate research, which have occupied our group at Imperial College, are:

(i) The ozone hole and the physical state of the stratosphere;
(ii) Global warming and the greenhouse effect.

We will now go on to consider, in turn, work on these two problems.

The Trouble with Ozone

Ozone (O_3) is a gas made up of three oxygen atoms. As such, its photochemistry is closely connected to that of molecular oxygen (O_2) through a basic series of reactions known as the Chapman reactions (Brasseur & Solomon, 1984):

$$O_2 + h\nu \rightarrow O + O$$
$$O + O_2 + M \rightarrow O_3 + M$$
$$O_3 + h\nu \rightarrow O_2 + O$$
$$O_3 + O \rightarrow O_2 + O_2$$

Because it requires the existence of a free oxygen atom (an unstable species), ozone is easily formed in the stratosphere, where high levels of short-wave UV radiation exist, energetic enough to split O_2 to form O. The stratospheric ozone layer exists between about 16 km and 40 km, with densities peaking in the 20 to 25 km range. The ozone layer absorbs solar radiation intensely between about 200 nm and 300 nm, thus protecting the Earth's surface from this highly energetic, and highly damaging (to biological systems) radiation. The ozone layer is, however, very tenuous: compressed to surface conditions, the thickness of the layer is a mere 0.3 cm. Hence our concerns over its vulnerability to chemical pollution, which might destroy ozone.

Ozone can be destroyed in catalytic cycles involving a reactive molecule (for example, Cl from fluorochlorocarbons (FCCs) or nitrogen dioxide (NO_x) from combustion engine exhausts). Typical catalytic reactions follow the scheme

$$\begin{array}{lcl} X + O_3 & \rightarrow & XO + O_2 \\ XO + O & \rightarrow & X + O_2 \\ \hline \text{Nett}: \ O_3 + O & \rightarrow & 2O_2 \end{array}$$

The species X can be Cl, NO, OH or Br, all of which can catalytically destroy ozone. The nett effect is to remove "odd oxygen" (that is, ozone or atomic oxygen) and convert it to molecular oxygen.

In the mid-1980s, Joe Farman, working at the British Antarctic Survey, recognised from a long period of careful observations that the ozone over the Antarctic was being depleted every Austral spring (Farman *et al.*, 1985). This was totally unexpected and caught the theoreticians completely by surprise. Eventually, after considerable work and controversy, this was ascribed to the effects of chlorine released from FCCs, reacting during the polar night at very low temperatures, on the surfaces of ice crystals (known as polar stratospheric clouds). The ozone depletion was occurring in October and November each year in the stratosphere, thus causing a roughly circular "hole" in the ozone layer over the Antarctic continent. This was clearly a cause of considerable concern. It forced governments to take the ozone problem seriously, and to actually get down to serious negotiations about limiting the emission of chlorinated agents into the atmosphere.

Several years prior to Farman's discovery, in 1978 in fact, NASA had started planning the Upper Atmosphere Research Satellite (UARS) which was to be an observatory in orbit around the Earth, to observe and study the stratosphere. Now the importance of this mission was emphasised by the new concerns about the ozone layer. UARS was launched in September 1991, carrying a suite of advanced Earth observation instruments designed to study the chemistry, dynamics and energetics of the stratosphere and mesosphere (Reber *et al.*, 1993). One of the instruments carried on board was the Halogen Occultation Experiment (HALOE) developed by NASA's Langley Research Centre in Virginia, and supported by an international Science Team, which included the present author. HALOE is a beautiful and sophisticated instrument that impressively is still operating as these words are being edited (August 1999): it measures the absorption of solar radiation by atmospheric trace gases as the sun sets or rises behind the limb of the atmosphere (Russell *et al.*, 1993). Analysis of the signals allows the density (or more usually, the relative density or mixing ratio) to be determined globally for a number of gases, as well as some other parameters: HF, HCl, NO, NO_2, H_2O, CH_4, O_3, aerosols and temperature.

Our group at Imperial College has particularly been concerned with using the measurements of water vapour, ozone and aerosols from HALOE to study the atmosphere (Harries *et al.*, 1996). We have used the data to investigate a number of phenomena, including:

(i) The dynamics of the stratosphere, mesosphere and upper troposphere;
(ii) The exchange of the material between the troposphere and the stratosphere;
(iii) The properties of the Antarctic polar vortex;
(iv) The effects of the Pinatubo volcanic eruption on the stratosphere (Pinatubo erupted in June 1991, just a few months before UARS was launched).

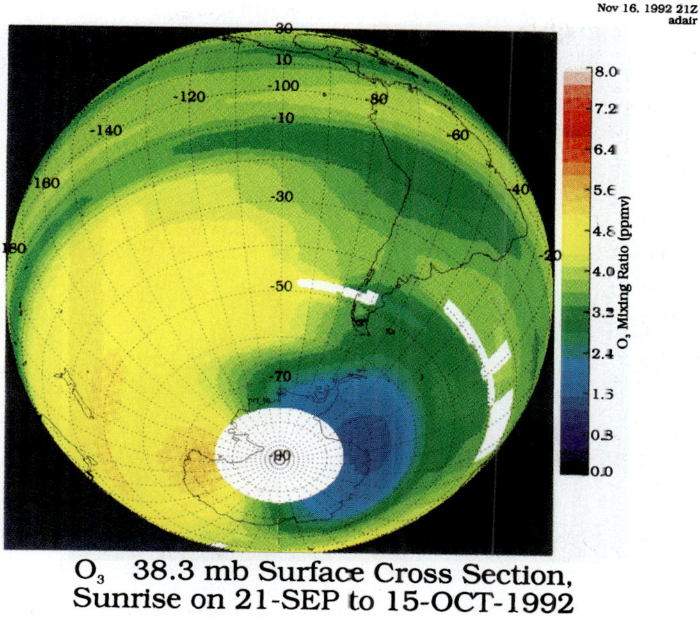

Fig. 3. HALOE observations of the ozone mixing ratio on the 39 hPa pressure suface between 21 September and 15 October 1992 (Courtesy of HALOE Project, NASA Langley).

The following illustrates just a small fraction of the results that have been obtained and published by colleagues and myself. Figure 3 shows one of many observations by HALOE of the Antarctic ozone hole: it shows the ozone mixing ratio measured on a pressure surface of 38.3 hPa averaged over the period 21 September to 15 October 1992. The reduced mixing ratios observed near the pole are very clear. Using many observations like this, over several years, and at many different pressure levels, allows us to study the mechanisms causing the growth and decay of the ozone hole. Figure 4 is a height-latitude cross-section, in which data have been longitudinally (zonally) averaged. The data shown are the water vapour mixing ratios in the stratosphere and mesosphere, clearly showing dessication occurring in several regions, especially over the Antarctic

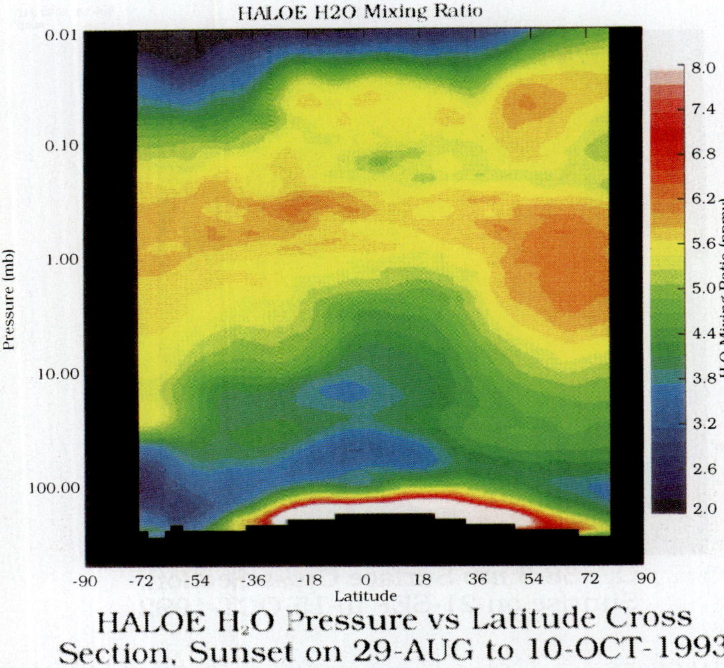

Fig. 4. HALOE observations of the water vapour mixing ratio shown on a pressure altitude versus the latitude plot (data are zonally averaged). Data taken over the period 29 August to 10 October 1993 (Courtesy of HALOE Project, NASA Langley).

(due to the formation of ice clouds), and in the mesosphere (due to the conversion of H_2O to H_2). Finally, Fig. 5 shows the extinction coefficient (a measure of the absorption of IR radiation) due to aerosols at a wavelength of 3.46 μm. The axes are pressure-altitude against time. The contours of extinction coefficient show the dramatic decay of aerosol with time after the Pinatubo eruption: this eruption was severe enough that a very large quantity of material was injected into the stratosphere, at altitudes of up to about 28 km, which circulated globally during the years following the eruption and decayed slowly. Measurements of decay rates give us information about transport within the stratosphere;

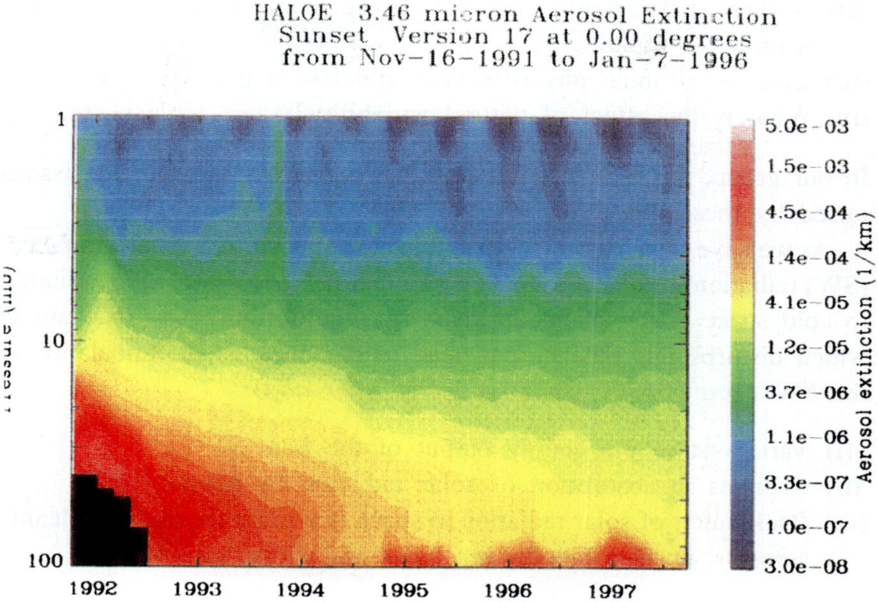

Fig. 5. HALOE observations at the equator of aerosol extinction coefficient shown as a function of pressure altitude and time. Measurements made at a wavelength of 3.46 μm from November 1991 to January 1996 (Courtesy of Chan, 1996).

measurements of the radiative absorption/emission properties of the aerosol tell us something about the climatic impact of the volcanic eruption.

A Comfortable Temperature in the Global Greenhouse

The background to the problem of the greenhouse effect and global warming is well known, and so does not need repeating here (see IPCC, 1996). Some of the current "hot" issues in climate research include:

(i) What are the principal mechanisms controlling climate and climate change?;

(ii) Is there a detectable trend in surface temperature due to global warming?;
(iii) Can we attribute any detectable trend to a particular cause?;
(iv) What is the effect of natural variability?

In our group, we have been attempting to address some of the issues raised by these questions.

As we have seen earlier (for example Fig. 2), incoming solar shortwave (SW) radiation is balanced by outgoing thermal longwave (LW) radiation to cold space if the planet is in energy equilibrium. Any mechanism which disturbs this balance can lead to climate change. Imbalance in the SW energy input can arise from (*inter alia*):

(i) Variations in the energy output of the Sun;
(ii) Changes in absorption of solar radiation by ozone;
(iii) Backscatter of solar radiation to space (for example, due to volcanic aerosols, surface albedo variations and cloud variations).

Imbalance in the LW energy output can arise from (*inter alia*):

(i) Variations in surface temperature or emissivity;
(ii) Changes in atmospheric absorption (for example, due to CO_2 or H_2O increases);
(iii) Variations in cloud amount, height or type;
(iv) Changes in atmospheric dust or aerosol loading.

Our work at Imperial College is focussing on the importance of changes in water vapour concentrations in the atmosphere; the radiative cooling to space from the upper troposphere via the pure rotation band of water vapour; and the SW and LW effects of clouds and aerosols.

To illustrate some of this work, Fig. 6 shows a calculation using a theoretical so-called "band model" of radiative transfer (Sinha & Harries, 1995). It shows a plot of radiative energy loss (the "cooling rate") due purely to water vapour, shown on a diagram of pressure-altitude against spectral wavenumber. The high intensities indicate regions where the cooling to space is most efficient, including the upper troposphere, radiating

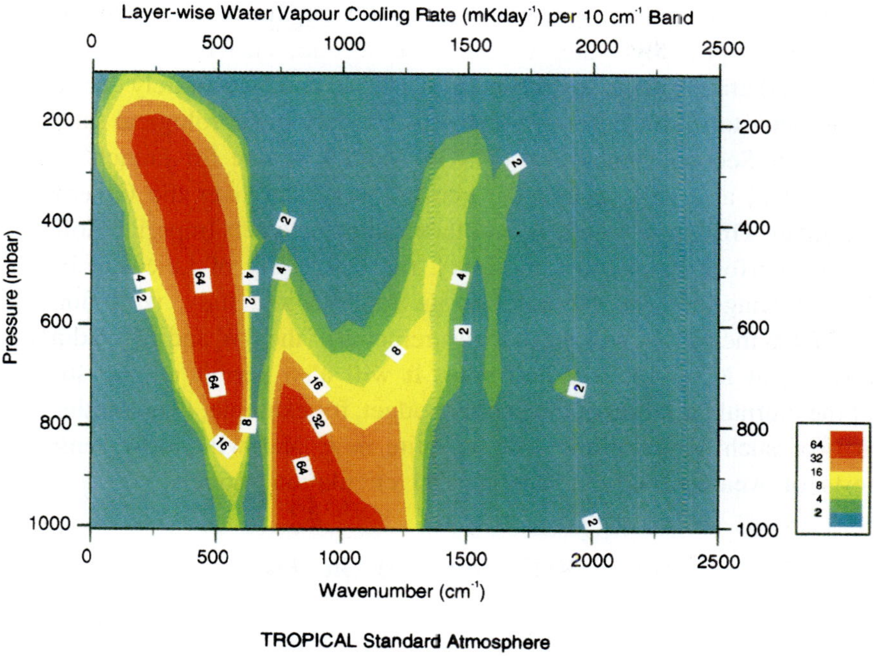

Fig. 6. Calculations using radiative transfer model of the heating rate (actually negative, hence a measure of rate of energy loss by radiation) for a tropical model atmosphere, and for water vapour only, expressed as a function of pressure altitude and spectral wave number (Courtesy of Sinha, 1995).

in the far IR from the pure rotation band. Until recent work by Clough and co-workers, and follow-up work at Imperial College, this strong cooling to space from the upper troposphere was not expected. It is an important planetary cooling mechanism for the maintenance of climatic balance.

Two other projects in this area are both experimental in nature. The first is called TAFTS (Tropospheric Airborne Fourier Transform Spectrometer). This is an accurate aircraft-borne radiometer for making direct measurements of spectral fluxes in the far IR within the atmosphere. Such measurements, at different levels within the troposphere and into the lower stratosphere, will allow us to test our theories of greenhouse cooling to space by water vapour. The second, GERB (Geostationary Earth Radiation Budget) experiment, is a broad band, accurately calibrated radiometer to fly on Europe's next geostationary meteorological satellite, Meteosat Second Generation (MSG). MSG-1 is due for launch in the year 2000, and will occupy the geostationary slot over the Greenwich meridian. GERB — which is a collaboration between Imperial College, the Rutherford Appleton Laboratory, the University of Leicester in the United Kingdom and laboratories in Belgium and Italy, with funding also from the EU — will make measurements of the SW and LW radiation arriving at MSG-1 every half-hour. It will also permit unique studies of the diurnal variation of the energy budget, for example, over continental Africa: such observations will be quite new, and will allow sensitive tests of weather and climate general circulation models.

Some Reflections on the Future

Status report

A new group has been established at Imperial College, based in the Physics Department, active in atmospheric and climate physics, using Earth observation from space, coupled with the application of powerful computer programmes for simulating the radiative, chemical and dynamical properties of the atmosphere. The new group is working on a variety of theoretical and experimental problems in climate and atmospheric physics, some of which have been illustrated above.

Scientific quality and relevance

Many difficult but fascinating problems still exist in trying to understand the Earth's climate system. This remains one of the greatest challenges in modern science — one of the "grand challenges" as we move into the third millennium. Not only is the subject one of great intellectual challenge, but it is also of enormous practical importance. The relevance of understanding our climate system to the peoples of this Earth is fundamental, for weather prediction, agriculture and food production, fresh water supply, coastline changes (sea level change), and for the whole stability of the global economic and political system which can all too easily be de-stabilised by even a small shift of climate patterns, precipitation, water supply and food production.

Controversy

Some observers have tried to undermine climate change research by suggesting that it is either unnecessary or a fabrication by scientists desperate for research funding. (Not all these observers are funded by interested groups, such as the fossil fuel lobby.) While it is extremely healthy to have criticism and controversy, so that a science does not become ossified, it is simply a fact that mankind must understand his climate system, and how his population numbers and activities relate to it. The Earth is the only survival system which we have available, and out of self-interest, if not out of altruism, we ought to understand and protect it.

Priority

The priority of climate research in the United Kingdom and elsewhere must be maintained. This is not a subject which decision-makers can treat as a passing fashion in science. In the universities in particular, the mechanisms for providing a degree of stability of research funding are urgently needed.

Conclusion

Understanding the basic physics of how our planet works, and how it might change in the future, remains a scientific problem of the highest priority and challenge to humanity.

This beautiful planet can be monitored and studied using the techniques of Earth Observation from space across all the colours of the rainbow and beyond.

Acknowledgements

My thanks for their originality, enthusiasm and friendship, and all credit for their contributions to this talk, go to my colleagues Drs Joanna Haigh and Ralf Toumi, our students, post-docs and other members of the group; to colleagues in the Blackett Laboratory and Imperial College; and to those who particularly helped with the preparation of this lecture, including Keith Clark and Nick Jackson.

References

Brasser, G. & Solomon, S. (1984) *Aeronomy of the Middle Atmosphere.* D. Reidel.

Chan, A. (1996) *Private communication.*

Farman, J.C., Gardiner, B.G. & Shanklin, J.D. (1985) Large losses of total ozone in Antarctica reveal seasonal CLO_x/NO_x interaction. *Nature*, **315**, 207–210.

Harries, J.E., Russell, J.M., Tuck, A.F., Gordley, L.L., Purcell, P., Stone, K., Bevilacqua, R.M., Gunson, M., Nedoluha, G. & Traub, W.A. (1996) Validation of measurements of water vapour from the Halogen Occultation Experiment (HALOE). *J. Geophys. Res.*, **101**, 10205–10216.

IPCC (1966) *Climate Change 1995.* Cambridge University Press.

Reber, C.A. (1993) The Upper Atmosphere Research Satellite (UARS). *Geophys. Res. Letts.*, **20**, 1215–1218.

Russell, J.M., Gordley, L.L., Park, J., Drayson, S.R., Tuck, A.F., Harries, J.E., Cicerone, R.J., Crutzen, P.J. & Frederick, J.E. (1993) The Halogen Occultation Experiment. *J. Geophys. Res.*, **98**, 10777–10797.

Sinha, A. & Harries, J.E. (1995) Water vapour and greenhouse trapping: the role of far infra red absorption. *Geophys. Res. Letts.*, **22**, 2147–2150.

CHAPTER 2

WATER IN A CHANGING WORLD

HOWARD S. WHEATER

Introduction

I have the honour of delivering the first inaugural lecture by a hydrologist at Imperial College. This is a recognition of the growing maturity of the discipline and the importance of hydrology for contemporary issues of environmental management. However, it belies a long and distinguished history. The College played a leading role in the emergence of hydrology as an academic discipline in the United Kingdom (UK) with the establishment, in 1955, of the first UK postgraduate course in hydrology. Over the next 40 years, staff and former students have made seminal contributions to research and practice, while, as will become apparent, hydrology has developed to become a multi-disciplinary science in its own right and the scope of hydrological applications has broadened dramatically.

This lecture aims to illustrate the development of hydrology and current issues of research and application by drawing on work in the hydrology research group at Imperial College. Basic hydrological processes are explained, and the importance of their understanding in resolving problems of environmental management is illustrated. To study processes in detail, small-scale observations are generally necessary, perhaps at the scale of an experimental soil column or a field plot of the order of tens or hundreds of square metres. As we move to larger scales, such as river basins or groundwater aquifers, the problem of integrating small-scale process studies to provide effective management tools becomes severe, and tensions between integrated modelling and empirical data analysis are reviewed in the light of current research. New developments are required to resolve important contemporary problems, such as global modelling; promising research directions are described.

As we shall see, hydrology has long been concerned with the prediction of the impact of change, but as we approach the twenty-first century, we are increasingly aware of the global scale of the issues of environmental management, and of their potential significance for human society and life on the planet. Hence the title of this lecture: "Water in a Changing World". I shall seek to illustrate the response of hydrology as a discipline to the increasingly complex questions posed by society and will comment briefly on the current political tensions which overlie a number of issues on water and environmental management.

What Is Hydrology?

The definition of hydrology

To take a contemporary definition (British Hydrological Society, 1994),

> "Hydrology is concerned with the processes governing the occurrence, distribution and movement of water in all its forms on, above and beneath the earth's surface. At a global scale, these processes can be represented as the hydrological cycle".

Hydrological processes

What are these processes? Figure 1 illustrates the terrestrial components of the hydrological cycle. Precipitation is the primary input, but may be intercepted by vegetation before reaching the ground. It may fall through the vegetation canopy (throughfall), run down the stems (stemflow) or be evaporated from the vegetation surface. At the ground surface, the water may run off or infiltrate into the soil.

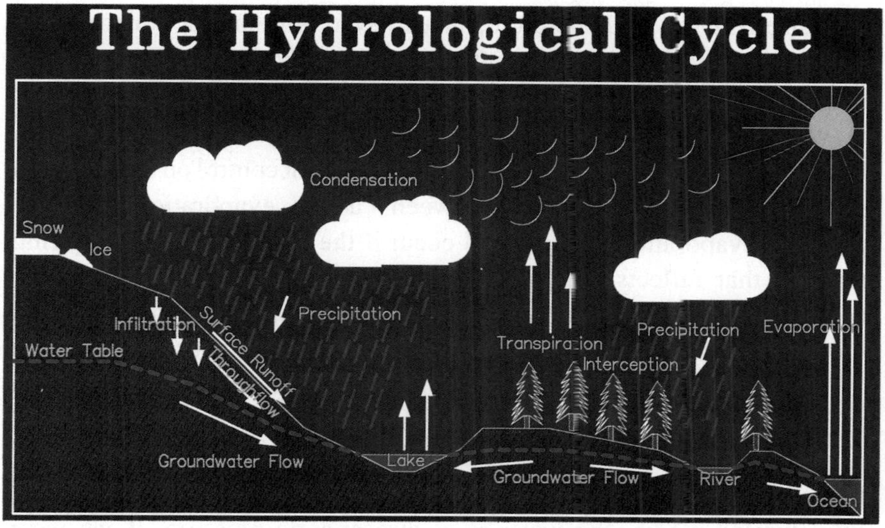

Fig. 1. The hydrological cycle.

The soil zone is complex; soil pores are generally partly air-filled (the unsaturated zone), and the resulting surface tension effects lead to soil water being held in the soil, rather like a sponge, under pressure less than that of atmospheric (soil suction). The drier the soil, the greater the suction is. Hence infiltration is controlled not only by gravity (which induces a downward flow), but also by these suction effects. Thus, in general, infiltration rates are higher for dry soils; as a soil wets, its infiltration capacity will decrease. Infiltration is also affected by surface conditions. Vegetation protects the soil surface from the energy of raindrop impact, which can cause crusting, as well as erosion. If rainfall intensities exceed the capacity of the soil to absorb water, overland flow will occur. For most natural surfaces in temperate climates, overland flow rarely occurs and/or is localised. This is not the case for bare soils in arid climates or paved surfaces.

The soil is the source of water for transpiration by vegetation. Most plants require some air around their roots (unsaturated conditions) to function effectively and, therefore, have to overcome soil suction to extract water. This they can do very effectively, but if the soil becomes too dry and the suction forces too great, the plants will be unable to extract the water they need. The first effect is a reduction in transpiration. Ultimately, the plants will wilt or even die. This plant control on transpiration leads to a possible difference between "actual evaporation" and the "potential evaporation", which will occur if the plant is freely transpiring at a rate that reflects atmospheric conditions. If the plant is stressed, for example, by water limitation, the actual evaporation rate will fall below the potential rate as the plant restricts water loss.

From the unsaturated soil, water may percolate downwards to reach the groundwater (the saturated zone). Where soils or rocks are saturated, the pores are full of water and surface tension effects are not present. Hence no suction forces arise. At the boundary between unsaturated and saturated media, the water is at atmospheric pressure. If we were to dig a hole, we would see this boundary as the "water table" — a free water surface. Below the water table, water in the pore space is under positive pressure due to its own weight. It will move under the effects of pressure and gravity as a groundwater flow — discharging at springs (where the water table meets the ground surface), or sustaining low flows in rivers, or discharging into lakes, seas or oceans.[1]

Alternatively, the complexity of near-surface soils may lead to shallow downslope flow (throughflow) via more rapid flow routes, whilst the effects of topography may further complicate the flow. Topographic convergence and downslope drainage lead to the establishment of saturated areas which can expand and contract as a catchment wets and dries, and which can contribute overland flow to stream response (Kirkby, 1978). Thus, we see a variety of pathways from rainfall to stream flow and an associated range of travel times (see for example, Wheater *et al.*, 1991a, 1993a). These are often, for the sake of convenience,

[1]If the water table is close to the surface, capillary effects can draw water up from the water table to the soil surface, where it can evaporate, leaving the salts it brings with it behind. This accumulation of salts is a highly damaging consequence of poor drainage in irrigated agriculture.

represented as two-flow components. Fast stream response — "stormflow" — arises from overland flow or near-surface pathways; slow "baseflow" arises from slower sub-surface drainage and groundwater discharge (Wheater *et al.*, 1982).

In the UK, an important seasonality in groundwater response occurs. In spring and summer, the evaporation rates usually exceed rainfall and plants draw on soil water to maintain transpiration. The soil water storage is depleted and a "soil moisture deficit" occurs. For significant groundwater recharge to take place, the soils must first rewet. This normally occurs in autumn and early winter. Thus, it is winter rainfall which is the key to next year's groundwater supplies.

Hydrological applications

These, then, are the processes. So what of their importance? The BHS document goes on to define the relevance of hydrology:

> "The hydrological cycle sustains the global climate and the global ecosystem, and hydrological processes determine the physical and biological environment. Water is essential for life: its distribution and occurrence is a major control on ecosystems and human society. It provides a transport medium for minerals, nutrients and pollutants — hence hydrological processes underpin the key biogeochemical cycles and frequently determine the fate of human pollution. It is the principal agent of erosion and landscape evolution".

These are, indeed, large issues in terms of their scope, scale and contemporary significance.

This modern view differs greatly from the early days of hydrology as studied, for example, by that first hydrology class of 1955. Historically, applied hydrology has developed to meet the needs of society. These have grown progressively more complex with the succeeding decades (Fig. 2). The main focus in the 1950s and 1960s was on the quantification of surface water and groundwater systems for water resources development

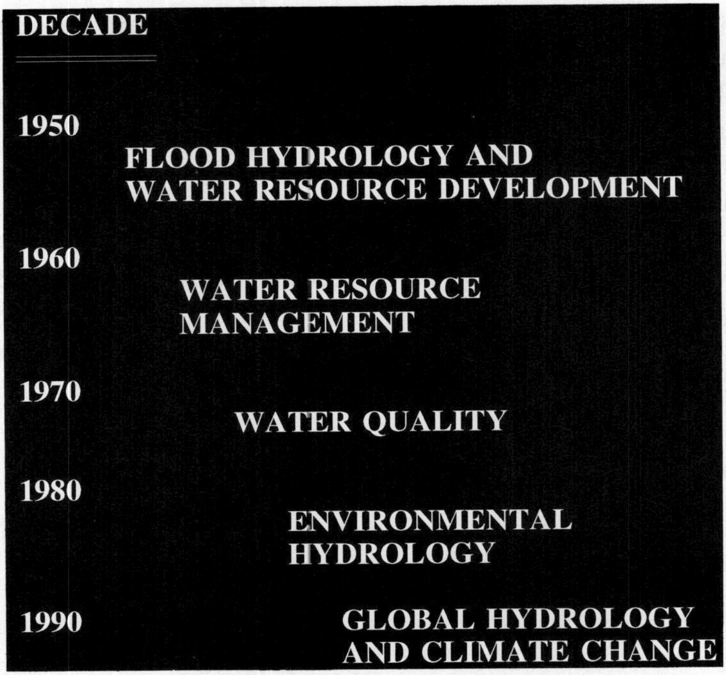

Fig. 2. Historical development of applied hydrology.

and flood protection. This was to meet the primary social needs of water supply and security from floods. This was followed by attention to the integrated management of water resource systems. By the 1970s, water *quality* had become an increasingly important constraint on water resource management. Thus, river water quality models were developed for design and operational management. In the 1980s, increasing concern with broader environmental issues was triggered by complex multi-disciplinary problems, such as the impact of acid deposits on surface waters and their associated aquatic ecosystems, as well as the fate of agricultural fertilisers. The acid rain problem transcended national boundaries and these problems had major political, as well as economic, implications. In the 1990s, global problems were perceived in the context of climate change and the hydrological community has become intimately involved with the issues of global climate — working, for the first time, on the hydrological cycle as a closed system.

The basic issues of water resources and floods remain a primary concern for much of the developing world, particularly as we move away from the humid, temperate climate conditions on which most hydrological methods (and textbooks) are based.

In the UK, basic problems remain, but set in a more complex framework. The sustainable development of water resources must balance abstractions from surface and groundwater sources against the environmental impact, such as the ecological effects of reduced river flows (possibly to zero) and the needs for dilution and assimilation of effluent discharges. Flood protection measures must also accommodate environmental impacts. An important concern for the £60 million Maidenhead-Windsor flood relief scheme, currently under construction, is the environmental impact associated with water quality in the constructed bypass channel. Elsewhere in Europe, differences in the political balance between the benefits of water development and the environmental impacts have been the focus of both internal differences and inter-state disputes. One notable example of this is the recent dispute between Hungary and Slovakia concerning the Gabcikovo-Nagymaros Barrage System on the Danube before the International Court of Justice in the Hague (see for example, Wheater, 1995).

Hydrological systems, thus, now appear in a much broader environmental context than ever before, and society requires quantitative answers to complex issues of environmental management, including the impacts of change.

Hydrological Process Research

What role, then, has hydrological science to play in the issues of environmental management? It is essential that we understand component processes if we are to predict their behaviour — particularly the impacts of change — and I will briefly illustrate this with three examples.

Forest interception

If we look back at the 1950s and 1960s, there was concern about the effects of afforestation of the British uplands. What would be the effect on the water balance and, hence, the yield of upland reservoirs? Field evidence from the UK and overseas showed conflicting evidence of catchment response. Two scientific developments were to resolve this.

The first was the development by Monteith (Monteith, 1965, 1973; Shuttleworth, 1993) of an earlier work by Penman (Penman, 1948, 1949) to include plant canopy effects in evaporation estimation. Penman had provided, arguably, the most important UK contribution to hydrology when he combined the aerodynamics of the transport of heat, water vapour and momentum from an evaporating surface with the surface energy balance to give an expression for potential evaporation (that is, representing the atmospheric controls on evaporation). This could be readily calculated from available data (temperature, humidity, wind speed and hours of bright sunshine). However, when transpiration occurs, evaporation takes place within the leaf and water vapour is lost to the atmosphere through the leaf stomata. Monteith explicitly represented this pathway by a stomatal resistance (or, at canopy-scale, a surface resistance): r_s. This allowed the effects of reduced stomatal aperture (for example, in response to limiting soil moisture) to be incorporated directly into a relationship for actual evaporation E:

$$\lambda E = \frac{\Delta R_n + \frac{\rho C_p}{r_A}(VPD)}{\Delta + \gamma\left(1 + \frac{r_s}{r_A}\right)},$$

where

λE is the energy of evaporation (λ is the latent heat).
R_n is the net evaporation.
VPD is the vapour pressure deficit.
Δ, ρ, C_p and γ are the temperature-dependent functions.

r_s is the stomatal resistance.
r_A is the aerodynamic resistance.

If the plant suffers water stress, it will respond by closing the stomata. Hence the resistance r_s will increase and the evaporation will decrease.

If we consider the evaporation of intercepted water, however, the stomatal pathway does not apply, since the water is evaporating from the leaf exterior. Hence

$$\frac{E_i}{E} = \frac{\Delta + \gamma\left(1 + \frac{r_s}{r_A}\right)}{\Delta + \gamma}$$

where E_i is the evaporation rate of the intercepted water.

Work at Imperial College by Rutter (1971, 1975) and elsewhere demonstrated that while for short crops the surface and aerodynamic resistance terms are approximately equal ($r_S \simeq r_A$), for forests the aerodynamic resistance may be an order of magnitude less ($r_A \ll r_s$), and hence, the evaporation rate of intercepted water is much greater than transpired water ($E_i \gg E$).

Given that the interception storage capacity is typically 1–2 mm, how significant, therefore, is this?

A simple computer model of these processes, the "Rutter model", can be used to demonstrate that in wet and windy climates, such as in Wales, the frequent filling and emptying of this store is of great significance, but that in dry East Anglia the effects are minimal. The data in Table 1, from the forested headwaters of the Severn, the adjacent grassland Wye catchment, and the Institute of Hydrology's experimental catchments at Plynlimon in central Wales, clearly show the practical effects of upland afforestation in terms of increased evaporation and reduced stream run-off.

This work is fundamental to the understanding of the impacts of land use change in the UK uplands and provides a general framework for evaluating the effects of land use change on the water balance. The

Table 1. Average annual water balance of the Plynlimon catchment.

Average Annual Water Balances, 1970–1980 (mm)

Catchment	Precipitation	Runoff	Evaporation
Severn (forest)	2293	1501	797
Wye (grass)	2417	2047	370

(After Calder, 1993)

application of this work continues, across a wide range of scales. For example, we are currently working with British Nuclear Fuels Ltd. (BNFL) at the Drigg low-level radioactive waste repository on the Cumbrian coast. From the measurement of canopy-level rainfall, throughfall and stemflow, we have calibrated a water balance model to quantify the effects of plantation forestry on the water balance of the Drigg site and surface water — groundwater interactions (Palmier, 1995; Burne *et al.*, 1997). On a very different scale, we are working with the Hadley Centre to validate schemes representing surface water and energy exchange in global climate models.

Radiochemical transport

Moving from water quantity to problems of water quality in vegetated soils, I turn to issues of waste management. These raise new questions for hydrologists and the nuclear industry (in particular, UK Nirex) is investing substantial resources in research to support the safety assessment of waste repositories. Radiological safety is concerned with human exposure

Plate 1. Lysimeter experiments in Silwood Park.

to radioactivity, and an important aspect is the transfer of contaminants from soils to vegetation (and hence to the food chain). Nirex recognised that current assessments of this aspect are simplistic, being based mainly on surface deposition data. What happens with contamination from below?

We formed a multi-disciplinary team at Imperial College and developed a set of lysimeter experiments at Silwood Park (Plate 1) to investigate the migration of a cocktail of radionuclides from a contaminated water table (Burne *et al.*, 1994; Wadey *et al.*, 1994). The data is comprehensive and unique, and include sub-surface fluxes, soil moisture, soil suction, meteorological variables, plant growth, root development and radionuclide concentrations in soil and vegetation. They have recently formed the basis for an international model inter-comparison study (Butler *et al.*, 1997).

The processes are complex. Plant transpiration develops an upward flux which advects the contaminant upwards from the water table, but

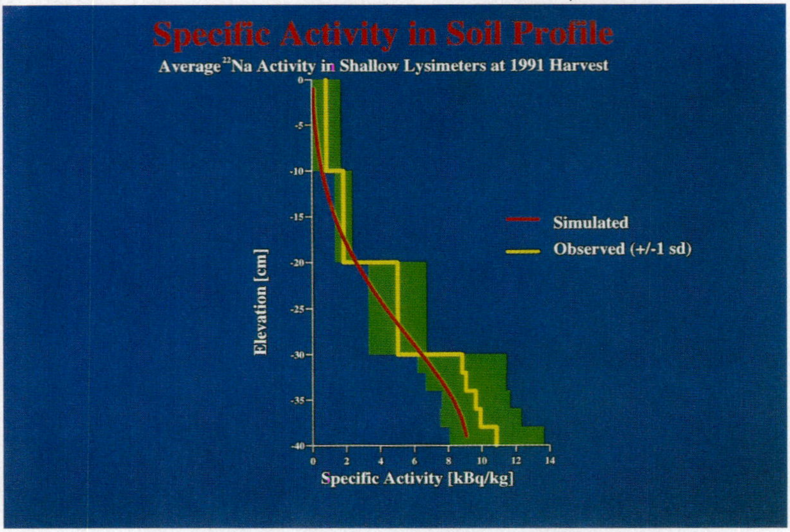

Fig. 3. Simulation of radionuclide concentrations in soil water.

this may be reversed by infiltration. The distribution of contaminants is, in addition, controlled by geochemical processes, particularly sorption. Plant uptake of contaminants depends on the root distribution and prevailing hydrological conditions, as well as local radiochemical concentrations.

We can represent these processes by two one-dimensional equations for the vertical transport of water and contaminants, including root uptake, sorption and radioactive decay. Figure 3 shows the simulated response of soil water concentrations of a mobile radionuclide to the annual cycle of upward migration due to high summer evaporation and downward flushing as winter drainage occurs. The performance of detailed modelling of the plant uptake from the various replicate lysimeter treatments is shown in Fig. 4 (calibrated on a single year). It shows that the essential characteristics of uptake and its variability are well-described. Finally, Fig. 5 shows a long-term simulation — using 30 years of detailed climate data — which illustrates a strongly nonlinear response to climatic

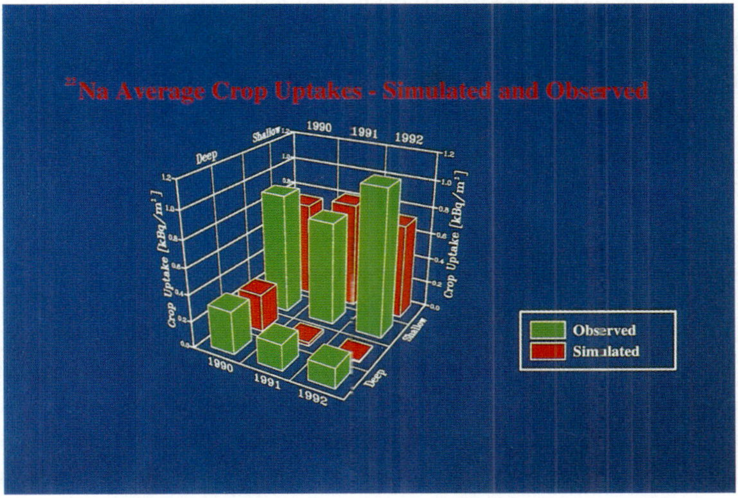

Fig. 4. Modelled and observed crop radionuclide uptake.

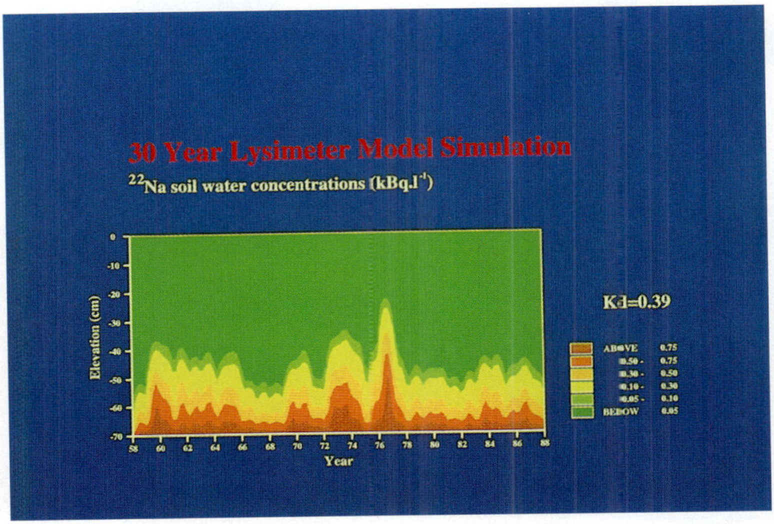

Fig. 5. Long-term simulation of crop radionuclide uptake.

variability (high uptake for the 1975/76 drought period) (Butler & Wheater, 1990).

It is apparent that previous attempts to represent soil-to-plant transfers, without recognising the hydrobiological controls, have been over-simplistic. On the basis of this work, more appropriate methods and parameters can be defined for improved safety assessment.

Transport of non-aqueous phase contaminants

If we move from vegetation and the soil to groundwater, we see other contemporary problems associated with contaminants. A wide range of problems of pollutant transport in groundwater exists, but some, in particular, challenge our current understanding of processes. For example, non-aqueous phase liquids (NAPLs) include petroleum products and organic solvents (such as trichloroethylene and perchloroethylene) that are highly toxic in concentrations exceeding parts per billion. They may be lighter

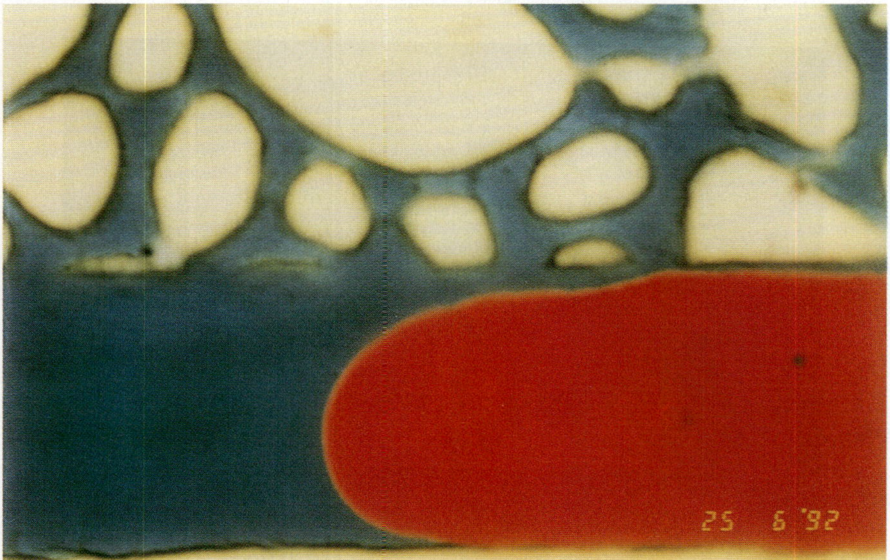

Plate 2. Micro-models for study of non-aqueous phase contaminants.

(LNAPL) or denser (DNAPL) than water. How, then, do they move, for example, in a fissured porous aquifer such as the chalk of southern England? Plate 2 illustrates results from laboratory-scale micro-models, using techniques developed in the Mineral Resources Engineering Department to investigate matrix penetration and flushing processes for DNAPL, as a precursor to model development (Zhou *et al.*, 1993; Zhou & Wheater, 1995). The plate shows the penetration of DNAPL, in this case carbon tetrachloride (in red), into a water-filled porous matrix (water in blue) as part of an investigation of the conditions for DNAPL penetration. To develop appropriate models to predict their movement and design remediation measures, we must have an improved knowledge of these complex flows.

Process Integration And Scale Dependence: Groundwater

The previous section demonstrates that an appropriate understanding of processes is essential to address a wide range of problems. However, for most practical purposes, our knowledge of processes must be integrated on a larger scale, for example, a river basin or a groundwater aquifer. We consider, first, groundwater systems.

Groundwater modelling

It has long been the case that groundwater models, based on a continuum representation of aquifer flow, have been applied to problems of groundwater management. The mathematical models are based on two simple concepts: the conservation of mass; and a relationship discovered by Darcy in the nineteenth century, namely, that flow in a porous medium is proportional to the applied gradient of potential energy (that is, due to the combined effect of gravity and pressure). The coefficient of proportionality is known as the hydraulic conductivity. Combining these two concepts yields a single equation (Bear, 1979) which can be readily solved in numerical models by the finite difference, finite element and boundary element methods.

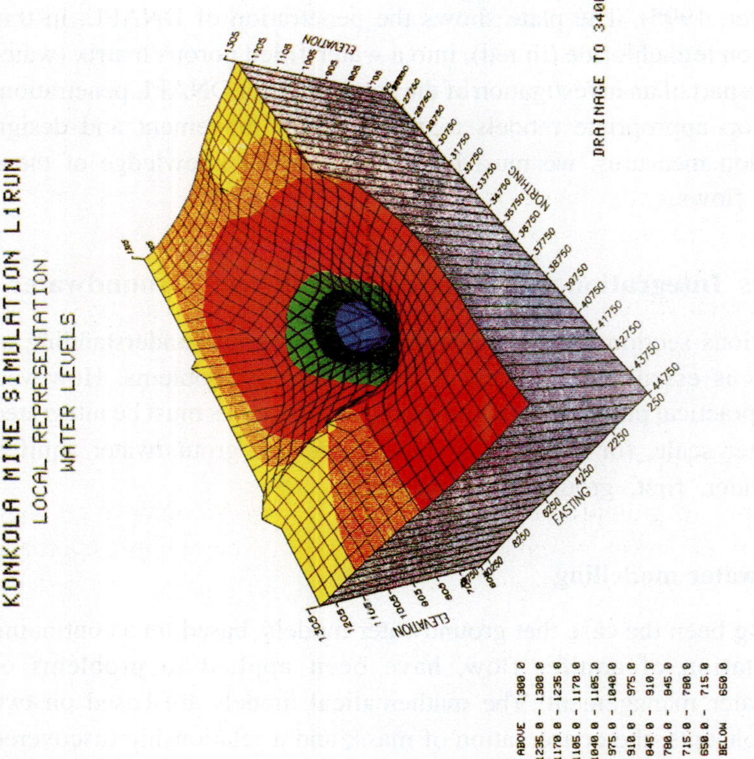

Fig. 6. Groundwater drawdown at the Konkola mine in Zambia.

Such models are powerful tools for a wide range of applied problems. In Fig. 6, we see the impact of the 1 km-deep Konkala Mine, one of Zambia's largest copper mines, on groundwater flow. Groundwater inflow from the river Kafue has resulted in a massive dewatering requirement, consuming some 3MW of power (Wheater & Butler, 1989).

Closer to home, we consider the problems of a proposed new town development on the Hampshire Chalk (Fig. 7). The model represents the system well (Fig. 8), but we note in passing that to fit the model to the data requires the specification of boundary conditions for the model (such as the flow at the lateral boundaries and groundwater recharge from overlying soils), as well as adjustment of the model's parameters which represent the physical properties of the Chalk aquifer, that is, the storage coefficient and the hydraulic conductivity. If the system is heterogeneous, this parameter adjustment usually involves subjective interpretation as the problem is essentially indeterminate (Wheater *et al.*, 1993b). Two parameters per grid square yields 500 degrees of freedom in fitting the model!

Groundwater quality modelling

If we move to groundwater quality, heterogeneity becomes a yet more important influence. As before, we can adopt a continuum approach, and hence, an advection-dispersion formulation for contaminant transport. The pollutants are advected by the groundwater flow through the pores and fissures of the medium, and dispersion (or spreading) occurs due to diffusion and small-scale mixing, represented by a dispersion coefficient (Bear, 1979).

For many problems, this formulation can work well. However, if we pool international experience, we see that our representation of the processes of dispersion is scale-dependent (Gelhar *et al.*, 1992). This is a failure of our underlying theory, which also presents practical problems. Contaminant migration, say, from a spill or leaking storage facility, may move from a local-scale to a regional-scale problem. How can we represent this?

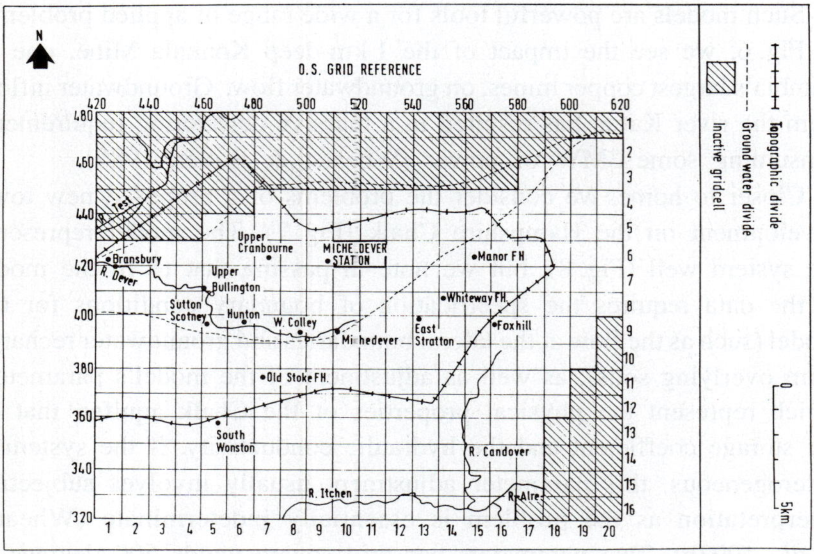

Fig. 7. Numerical grid for a groundwater model of the River Dever catchment in Hampshire.

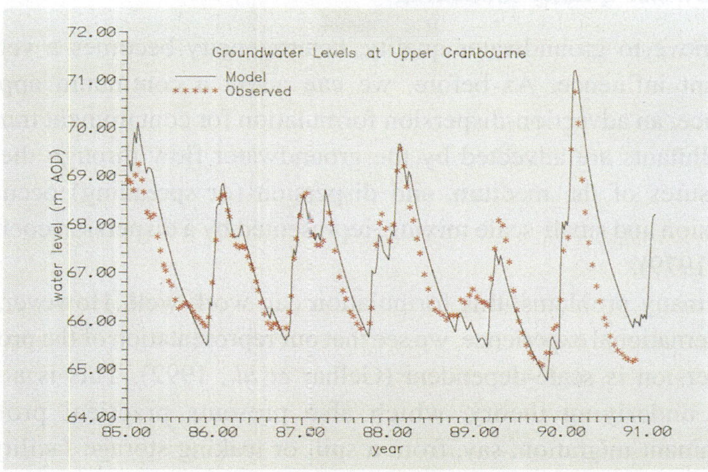

Fig. 8. Simulated and observed groundwater levels in the River Dever catchment, Hampshire.

The problem is essentially due to heterogeneity in the groundwater system (Dagan, 1989). If we include a deterministic representation of small-scale heterogeneity, we can reproduce this scale dependence, but we cannot hope to define this small-scale structure uniquely from available data. Hence we move to a geostatistical framework, where we can develop equally likely representations of the small-scale structure — constrained by observed measurements — and define a probability distribution of contaminant migration using Monte Carlo simulation (that is, repeated random trials) (Tompkins *et al.*, 1994). This state-of-the-art development takes groundwater flow and quality modelling forward to include a realistic definition of the large uncertainties in the prediction of groundwater pollutant transport, and to provide new tools for, for example, the appropriate definition of aquifer protection zones (Fig. 9) (Wheater *et al.*, 1998; van Leeuwen *et al.*, 1998a, 1998b).

Fissure flow

Problems become yet more severe when fissure flow predominates and our continuum representation breaks down. Again, we have insufficient information to specify fissure geometry and structure uniquely. We need new techniques to characterise such systems, particularly where NAPL is concerned.

At the Nordegg gas refinery in Alberta, Canada, we developed a research site for LNAPL investigation (Hardisty *et al.*, 1994). Here, a fissured sandstone has been transmitting gas condensate from a leaking disposal pit to the river (an example where US$6,000 of protective measures could have prevented a $6 million clean-up programme). Hardisty applied oil industry investigation methods to new insights into the transmission and entrapment of the LNAPL, and hence, to design and implement appropriate remediation measures. Unexpected things can happen. For example, it was shown that light contaminants — present in fissures and floating on the groundwater — can be pumped upslope by the natural variability of the groundwater surface — a further example of the importance of process understanding to the resolution of applied problems.

Fig. 9. Schematic representation of the construction of a stochastic well capture zone. (a) Three realisations of a random hydraulic conductivity field; (b) the capture zones associated with these fields; and (c) the stochastic capture zone showing contours of probability of capture.

Process Integration And Scale Dependence: Surface Water

We now turn to surface water and the response to rainfall of rivers and streams. As hinted at earlier, if we consider water movement in a natural hillslope, we see heterogeneity and complexity of response. Yet at catchment scales, these responses are integrated to give a smoothed streamflow hydrograph. The extent to which we can replicate this integration in modelling these rainfall-runoff processes is an important, unresolved issue in hydrology.

Rainfall-runoff modelling

At a simple level, we can represent the transformation between rainfall and streamflow at catchment scale by a volumetric loss and a time-distribution function. We can also identify the response *uniquely* from the data by methods of linear analysis (Wheater *et al.*, 1993b). This allows regional analysis, which enables one to define, empirically, the dependence between the model parameters that define the rainfall loss and flow transformation and catchment physical characteristics — the basis of flood event design methods throughout the world (NERC, 1975; Wheater *et al.*, 1982). However, this important feature of identifiability is associated with the constraint of a simple model structure. We have shown (Beck *et al.*, 1990) that the streamflow hydrograph, in general, contains only sufficient information to identify a very simple (two-component) model, which may be physically unconvincing and have only moderate performance.

For continuous simulation, it is common for hydrologists to incorporate a more complex and subjective interpretation of internal processes into a "conceptual model" which can be fitted to the observed response. The functional relationships in the model are defined by model parameters, which may have some physical significance and must, in practice, be adjusted (either manually or automatically) until the model adequately fits the calibration data. Such models, which are generally based on a prior perception of the dominant processes, are powerful tools routinely used for catchment management. However, problems of parameter identification were initially defined in work at Imperial College by Ibbitt and O'Donnell (1971) 25 years ago. We now recognise that, for most current models of this type, model parameters are not uniquely identifiable. The term "equifinality" has been coined (Beven & Binley, 1992) to represent the fact that parameters in such models are usually interdependent. Hence different combinations of parameter values can give very similar model performance. At catchment scale, the data available are insufficient to identify internal processes at this level of complexity.

For many applications, this may not matter. However, it means that predictions of the effects of catchment change can only be made with unspecified levels of uncertainty, and that regionalisation of models to allow application to ungauged catchments has, as yet, been largely unsuccessful. Current research is aiming to improve identifiability through new developments in parameter estimation methods (Wheater *et al.*, 1986; Kleissen *et al.*, 1997) and more efficient model structures (Wheater *et al.*, 1993b)

With currently available computing power, we can build a continuum-based model of surface and sub-surface processes based on our knowledge of the underlying physics. However, can we represent the effects of process heterogeneity in these models? Our experience of modelling and experimentation on forested hillslopes is that preferential flow paths are incompatible with a simple continuum model (Koide & Wheater, 1992). How can we specify the physically-based parameters at an appropriate scale for application to real catchments? At present, there is no adequate basis, and calibration using observed data remains essential, at least for sub-surface components of streamflow. This reinforces the problems of parameter identification and equifinality (Wheater *et al.*, 1993b).

These are difficult issues, but they are the subject of much current research. Can we aggregate from existing theory? Or do we need a new holistic approach? Figure 10 illustrates the strengths and weaknesses of the above approaches. One way forward is to develop a stochastic approach, analogous to that for groundwater, with model conditioning (for example,

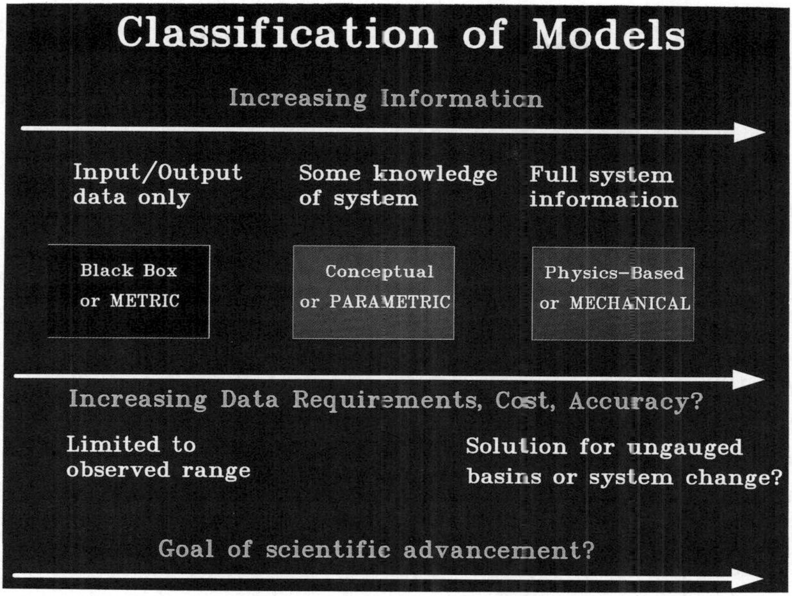

Fig. 10. A classification of rainfall-runoff models.

using remotely sensed data). Another is to combine the strengths of physics-based and data-based approaches in a hybrid methodology, which is the subject of ongoing research (Lees *et al.*, 1998).

Surface water acidification

These issues of hydrological modelling were brought into sharp focus by the problems of surface water acidification. The extent to which acid rain is neutralised depends on hydrological flow paths and their associated chemical changes (Mason, 1992; Wheater *et al.*, 1993a; Wheater & Beck, 1995). In upland catchments with acid soils and an underlying geology of slowly weathering rock, near-surface flows have little opportunity to neutralise acid inputs. Deeper flow paths interact with weathered material and acidity can be dramatically reduced by cation exchange and other chemical and biological processes. How can we identify this combined flow and chemical response? In an attempt to analyse catchment response, there has been a move from the (subjective) conceptual models — widely used to predict impacts of change — to simpler models, which can at least give a unique definition of response (Wheater & Beck, 1995; Beck *et al.*, 1990; Kleissen *et al.*, 1990).

In practice, we need improved process understanding, which can only come from painstaking field research. For example, we have been working for several years to define the hydrochemistry of a small (four hectares) tributary of the Wye at Plynlimon (Fig. 11) (Muscutt *et al.*, 1993; Chapman *et al.*, 1993; Hutchins *et al.*, 1995). The hydrology is complex, with extensive occurrence of natural pipes within the peat soils. We see sources of aluminium in the pipe walls, dramatic changes in pH, base cations along flow paths to the stream and in-stream biological processes which moderate stream chemistry (Fig. 12). We are presently working to identify the scale dependence of these responses in the Plynlimon catchments, and to develop new modelling approaches which combine appropriate representation of (uncertain) hydrological and geochemical processes with available spatial data of soils, topography and streamflow.

Although political pressures have moved elsewhere, the issues of acid rain remain a major scientific challenge. We still have much to learn about the combined processes of hydrology and geochemistry which determine the responses of our upland catchments to atmospheric deposition and land use change.

Arid zone hydrology

Thus far, we have considered processes and process aggregation independently of climate variability. At this point, I would like to return to the hydrology of arid zones — from Wales, where it rains three out of four days, to Arabia, where it may rain only three days in a year — to reinforce our previous discussion of the problems of catchment-scale analysis and, to add an additional element, rainfall, a primary determinant in catchment-scale response.

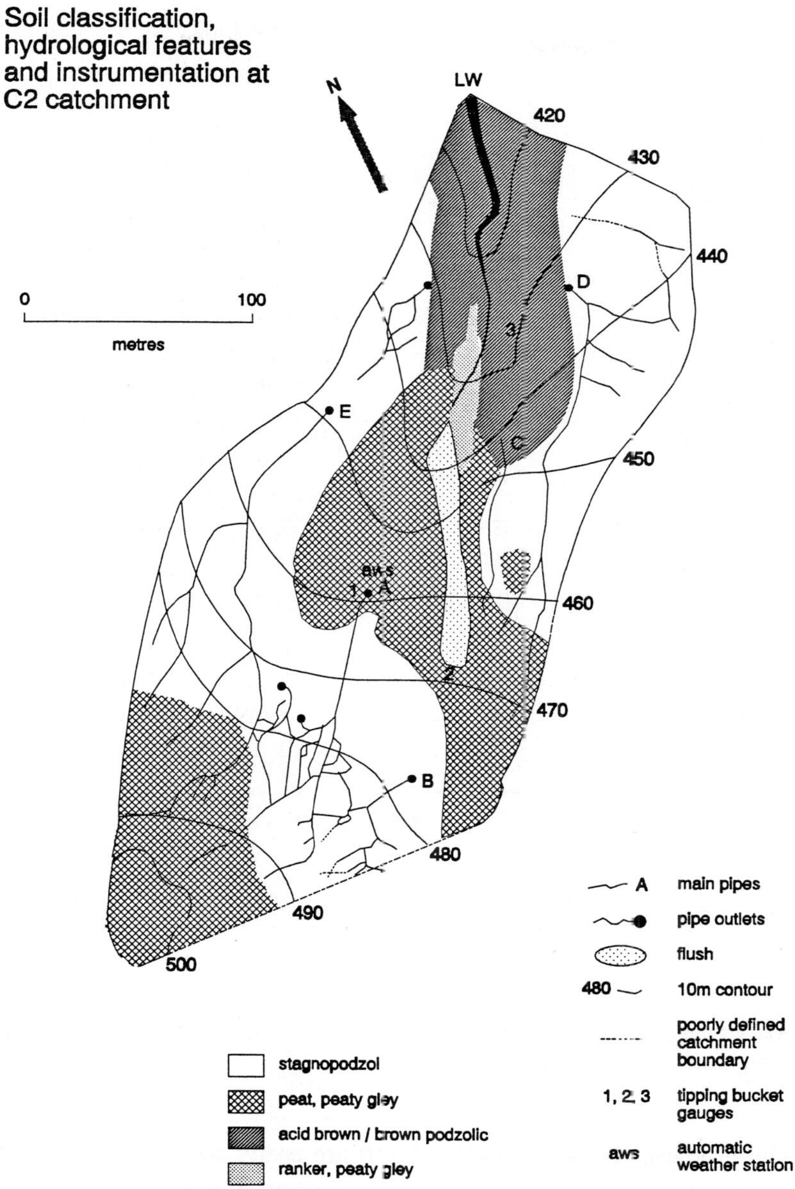

Fig. 11. The C2 catchment in Plynlimon, Wales.

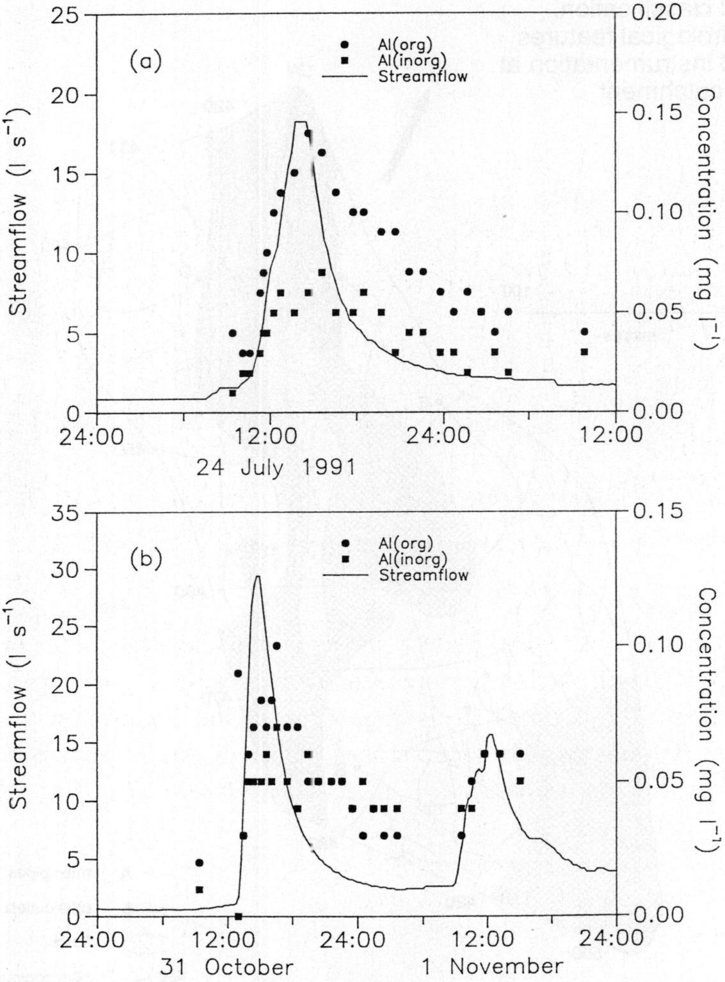

Fig. 12. Examples of storm response of organic and inorganic aluminium concentrations in stream water for the C2 catchment.

The problems of floods in the desert are easily overlooked. However, they are important and frequently life-threatening. Plate 3 shows all that remains of a newly completed road embankment following the floods

Plate 3. Remains of road embankment after the May 1981 flood, N. Oman.

of 3 May 1981 in Oman. Plate 4 shows a water level recorder in Wadi Yiba, S.W. Saudi Arabia. Its predecessor was washed away by a flood of 3000 m^3s^{-1} (30 times an average flow in the Thames) with a rise time of 30 minutes. The government of Saudi Arabia invested some £10 million in a comprehensive study of the hydrology of five basins in S.W. Saudi Arabia in the 1980s. The data have been of fundamental importance in improving our understanding of arid zone hydrology (Wheater *et al.*, 1989, 1991b, 1991c; Wheater & Brown, 1989).

Annual rainfall on the Red Sea coastal plain is around 50 mm, but in the mountains this may reach 450 mm at high elevation. Plate 5 shows a small flood moving down a dry wadi bed. We note a clear, almost cloudless sky, and the approach of sunset. Where did the water come from?

Dense rain gauge networks were installed and the rainfall characteristics are illustrated in Fig. 13, which shows the spatial pattern of daily rainfall for a typical event on Wadi Yiba and its hourly distribution. Clearly,

Plate 4. Water level recorder in Wadi Yiba, S.W. Saudi Arabia.

a rain gauge spacing of approximately 10 km is inadequate to capture the true spatial distribution. Locally intense rainfall generates overland flow on the bare soils of the area, which in turn generates a flood flow in the wadi channel. However, the bed of the channel is highly permeable. Hence the flood infiltrates the bed as it proceeds downstream and progressively reduces in magnitude. The bigger the flood, the further it is likely to travel.

We can clearly see that a measurement of wadi discharge is an incomplete indicator of catchment response. It cannot be used to infer the upstream

Plate 5. Flood flow in Wadi Yiba, S.W. Saudi Arabia.

processes. Indeed, zero flow does not imply that flows were not generated upstream. The spatial pattern of rainfall is a dominant influence, but is unlikely to be well-defined by conventional gauging methods. Wheater and Brown (1989) applied a conventional analysis of rainfall and streamflows and found that, apparently, the highest flood is generated by the smallest rainfall.

There are obviously important implications for the methodologies of flood design, but there are particular problems in the assessment of water resources. These flood flows provide infiltration which can sustain invaluable shallow groundwater resources. Important questions concern their potential for further exploitation and their long-term sustainability. We are currently attempting to define this for the sand rivers of botswana (Andersen & Wheater, 1998).

We have also recently been working in Oman to evaluate the impact of dams to retard the flood flows and, hence, locally increase groundwater recharge (Wheater *et al.*, 1995). We have had to confront the issues

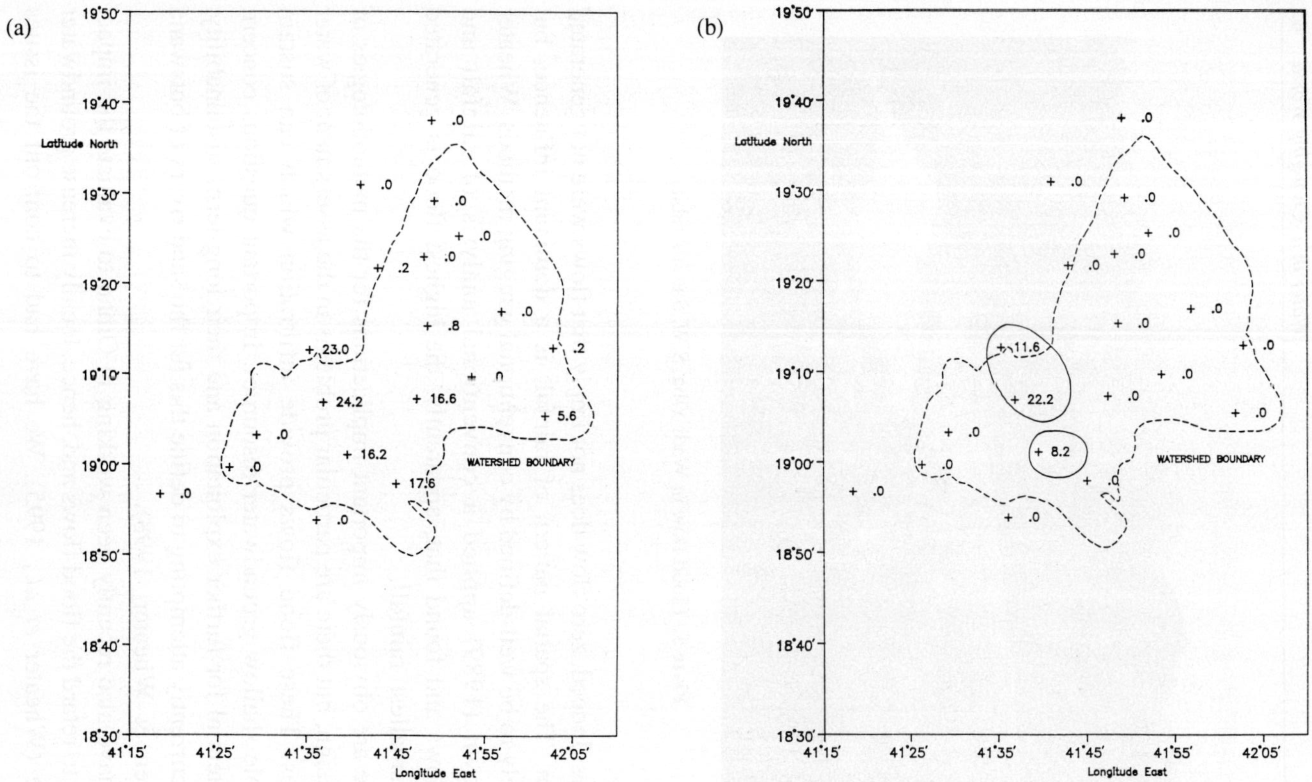

Fig. 13. Rainfall in Wadi Yiba, S.W. Saudi Arabia. (a) Daily rainfall totals on 19 September 1984; (b) the first hour of rain (1600 hrs); (c) the second hour of rain (1700 hrs); and (d) the final hour of rain (1800 hrs).

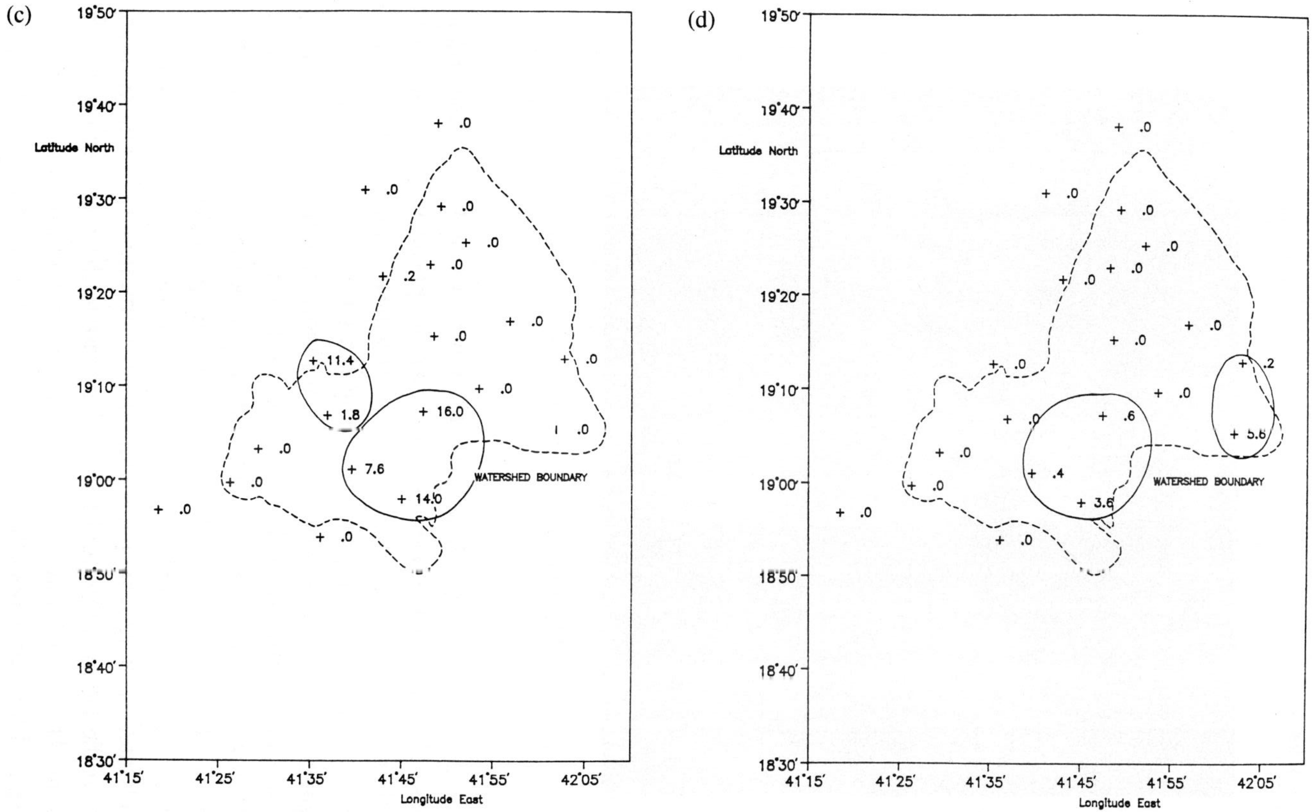

Fig. 13. (*Continued*).

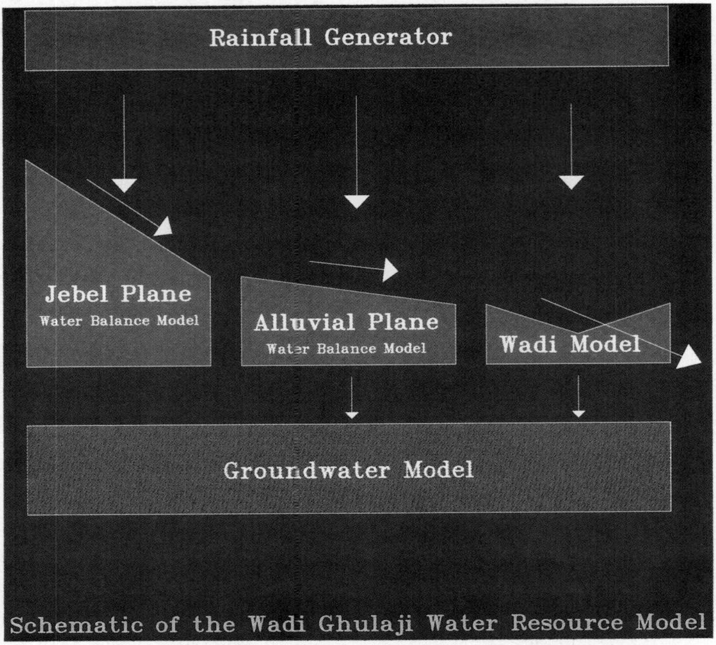

Fig. 14. Arid zone water resource simulation model.

of spatial variability, which require distributed modelling of the runoff process. The solution has been to develop a new approach to the distributed modelling of arid zone catchments. A stochastic multivariate rainfall model was developed to simulate long-term (100 years) data for the rain gauge network, as input to a rainfall runoff model, which generates wadi flows and infiltration (Fig. 14). The surface water model is coupled to a distributed groundwater model, and the performance of the integrated model for the few observed events available provided constraints on the model parameters. It also allowed the evaluation of options to manage groundwater recharge, in particular the construction of recharge dams to retard the flood flows, and hence, concentrate the recharge on desired locations. However, more work is needed to characterise spatial rainfall. New spatial-temporal modelling techniques are currently being investigated

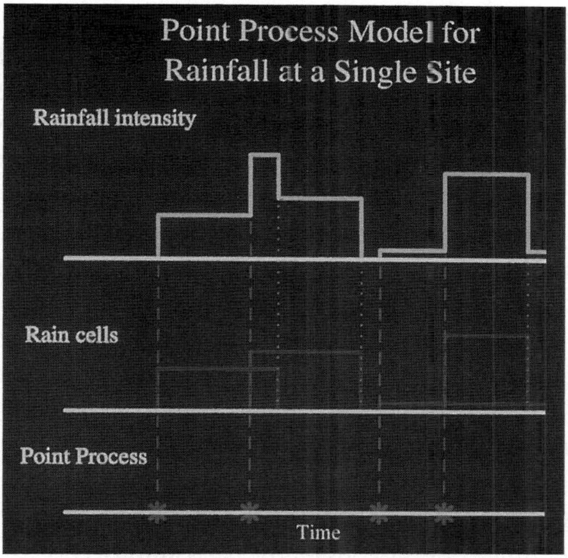

Fig. 15. Rectangular pulse rainfall model.

(Samuel, 1999) based on detailed data from the Walnut Gulch experimental area in Arizona (Osborn *et al.*, 1979).

Rainfall

The problems of the arid zone highlight the importance of rainfall in that regime, but rainfall is of fundamental importance to most aspects of hydrological response. In general, until relatively recently, rainfall has been a neglected area of research, particularly in the context of hydrological design. However, there are two important areas of contemporary development which received attention at Imperial College.

Cox and Isham have pioneered the theoretical development of point rainfall models based on Poisson cluster processes (Cox & Isham, 1980; Rodriguez-Iturbe *et al.*, 1987, 1988). The principles are simple and retain a physical conceptualisation. Storm arrivals occur as a stochastic process and then, within the storms, storm cell occurrence is generated. A simple cell representation is commonly adopted, such as a rectangular pulse in time (Fig. 15). Onof and Wheater have evaluated this family of models on British rainfall data (Onof & Wheater, 1993, 1994) with considerable success in reproducing point rainfall statistics across a range of temporal scales. The models have recently been adapted to generate (point) rainfall sequences which can be applied in a stochastic framework for hydrological design (Onof *et al.*, 1996a). Thus we can generate likely rainfall profiles and evaluate the resulting catchment response. This contrasts with conventional flood design practice where, for a rainstorm of a given duration, a storm depth is selected and synthetically distributed in time according to a symmetrically distributed pattern. There is, justifiably, increasing concern about the application of these traditional methods for systems that are sensitive to rainfall temporal distribution.

As we have seen, spatial rainfall distribution can be an important source of uncertainty. We also urgently need improved characterisation of the spatial-temporal rainfall process for a variety of hydrological applications. Weather radar is a technique which has provided new insights into that process, although not without important problems in the calibration of rainfall intensity. The Natural Environment Research Council has also been supporting a Hydrological Radar Experiment (HYREX) through the installation of a dense calibrating rain gauge network and the provision of coincident data from three radar systems. In collaboration with Isham, Cox and Rodriguez-Iturbe, we are investigating the spatial rainfall properties, including an evaluation of the extension of the Poisson cluster methods to the spatial-temporal process (Northrop, 1996; Wheater *et al.*, 1998a). The results are encouraging (Fig. 16) and offer the possibility of significant advances in hydrological design (Wheater *et al.*, 1997).

Global Modelling

A realistic representation of surface hydrology is essential in global models which attempt to simulate global climate and predict climate changes, whether natural or man-made. The vertical exchange of energy and water

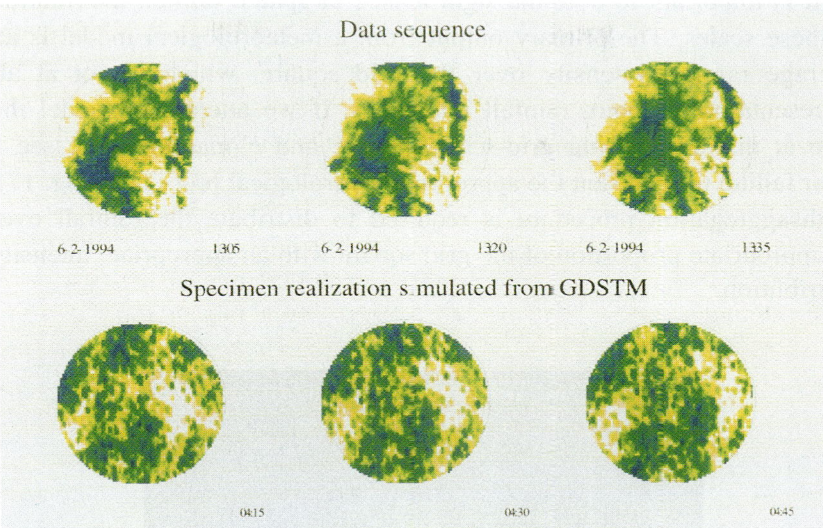

Fig. 16. Performance of a spatial-temporal rainfall model in comparison with a radar-observed sequence.

vapour at the earth's surface has an important impact on climate. However, river flows are also required to represent discharge into seas and oceans, as well as to major lakes and wetland system, such as the Sudd or Okovango. This raises the issues of how to represent the heterogeneity of surface hydrology and climate variability at very large scales. Thus, it combines the problems of model aggregation discussed above with the issues of spatial variability of climatological variables.

The current generation of GCMs has a grid element size of the order of 200 km × 300 km. However, the same issues arise for smaller scale meteorological models with grids that are as small as 10 km. How, therefore, should the sub-grid variability of soils, land use and hydrological processes be represented for global application? What are the major controls on hydrological response on a large scale? How can this be evaluated?

It is important to note the significance of spatial rainfall distribution at these scales. The primary output from a meteorological model is an average rainfall intensity over the grid square, which is not at all representative of point rainfall intensities. If we attempt to model the Nile at this scale using grid-scale average and climatology, we see a clear failure to represent the appropriate hydrological processes (Fig. 17). A disaggregation procedure is required to distribute the rainfall over an appropriate proportion of the grid square with an appropriate intensity distribution.

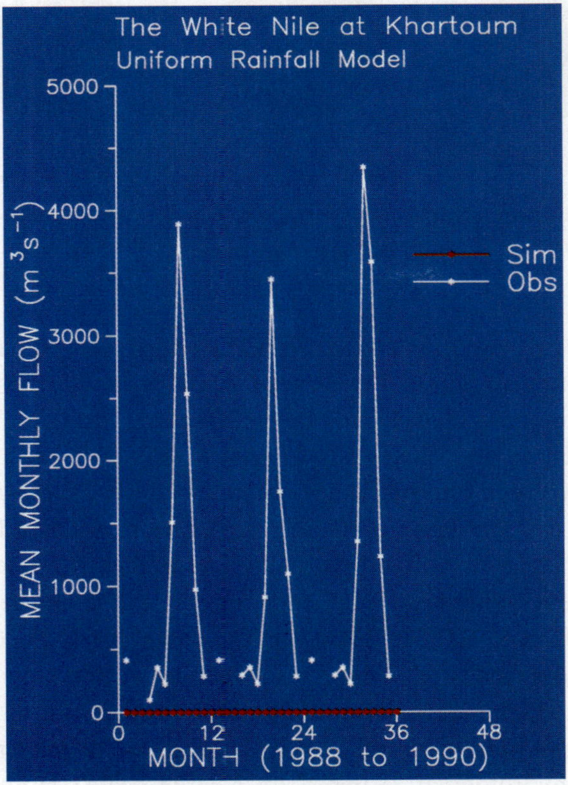

Fig. 17. Simulation of River Nile flows using grid-square averaged inputs.

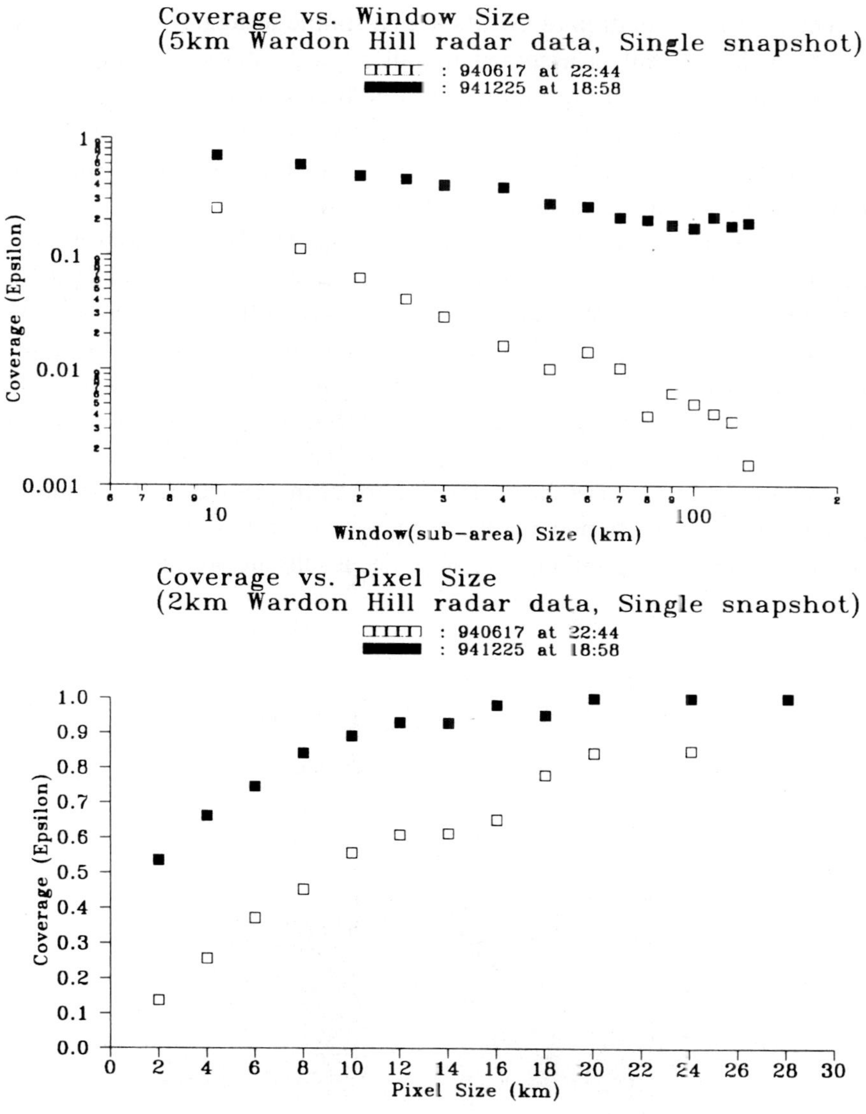

Fig. 18. Analysis of spatial coverage of rainfall. (a) Wardon Hill radar, 5 km resolution data. Rainfall coverage as a function of model grid element size; and (b) Wardon Hill radar, 2 km resolution data. Rainfall coverage as a function of scale of image resolution.

The current generation of GCMs uses a simple disaggregation scheme in which the rainfall is assumed to fall on a proportion of the GCM grid square (dependent on rainfall type) with an exponential distribution of intensity (Warrilow *et al.*, 1986). However, they do not have a strong basis in data analysis. They also have no spatial memory, from one time-step to the next, of where rain has fallen.

We have been working, with support from the NERC TIGER programme, to improve these methods through an analysis of the radar data from the UK (Onof & Wheater, 1996a, 1996b) and the Mississippi basin (Onof *et al.*, 1996b) (Fig. 18). Guidance has been provided to GCM modellers, with respect to simple improvements to the distribution-based disaggregation schemes. Current work (Wheater *et al.*, 1998b) has developed a new location-based disaggregation scheme, based on the Markov random field methods, to preserve the spatial pattern of the disaggregated rainfall, its memory — conditioned on the grid square rainfall intensity — and the distribution at the previous time step. The results are promising (Fig. 19)

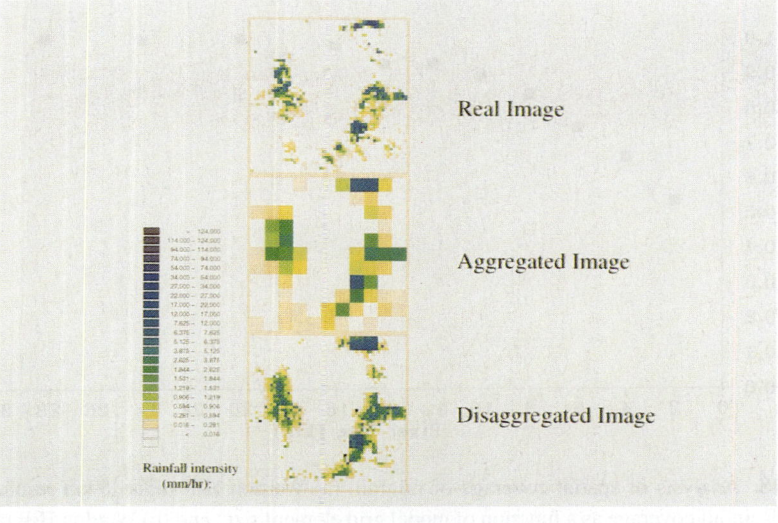

Fig. 19. Performance of Markov random field rainfall disaggregation procedure.

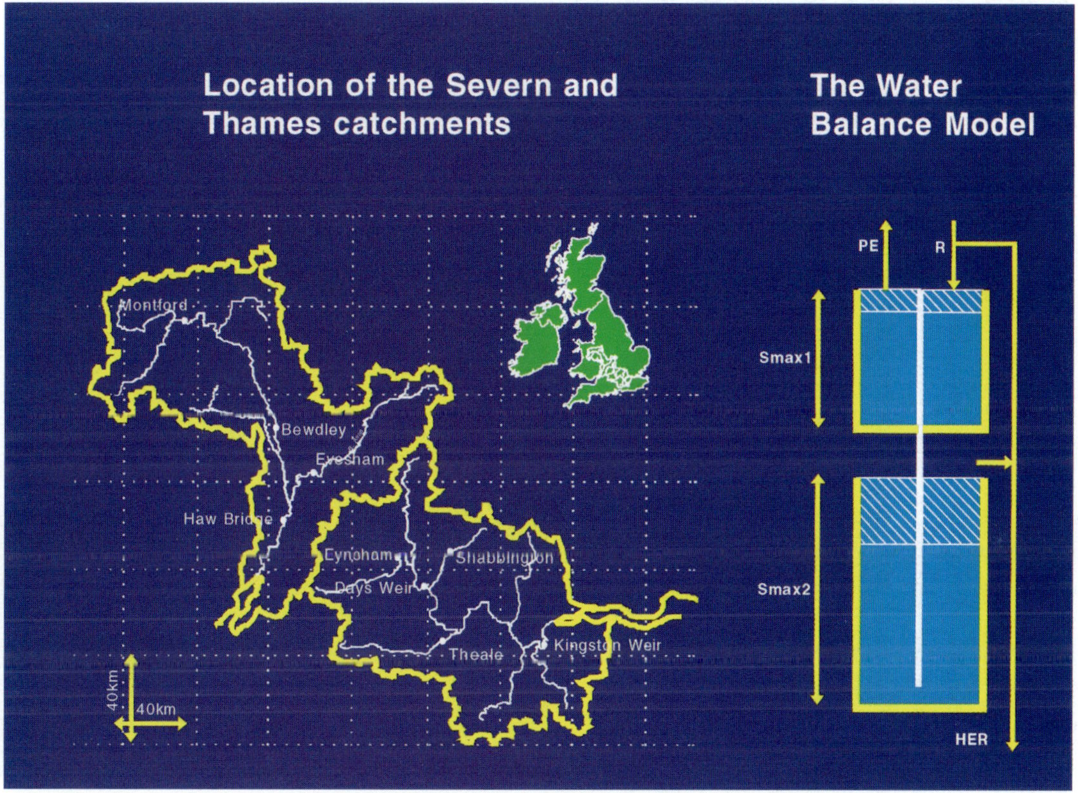

Fig. 20. Grid-based rainfall-runoff model for Rivers Severn and Thames.

and the method has wider applicability in meteorological modelling and short-term forecasting.

The performance of relatively simple grid-based hydrological models has been investigated for the Severn and Thames (Fig. 20). The impacts of climate on sensitivity to rainfall and soil moisture disaggregation have also been defined. Performance in reproducing the water balance of the Severn and the Thames has been impressive. Using a 40 km grid square resolution, and with land use distribution represented by averaging the response of 14 land use elements, the annual water balance was simulated, with no calibration, to within 0.5% for the Thames and 2.3% for the Severn (Jolley & Wheater, 1996). However, in order to reproduce the temporal dynamics, we need to incorporate a simple algorithm to represent sub-surfacing routing and the effects of the channel network (Fig. 21). Here, calibration of a single parameter per grid element is required (Jolley *et al.*, 1997; Wheater *et al.*, 1998b). Thus, we have

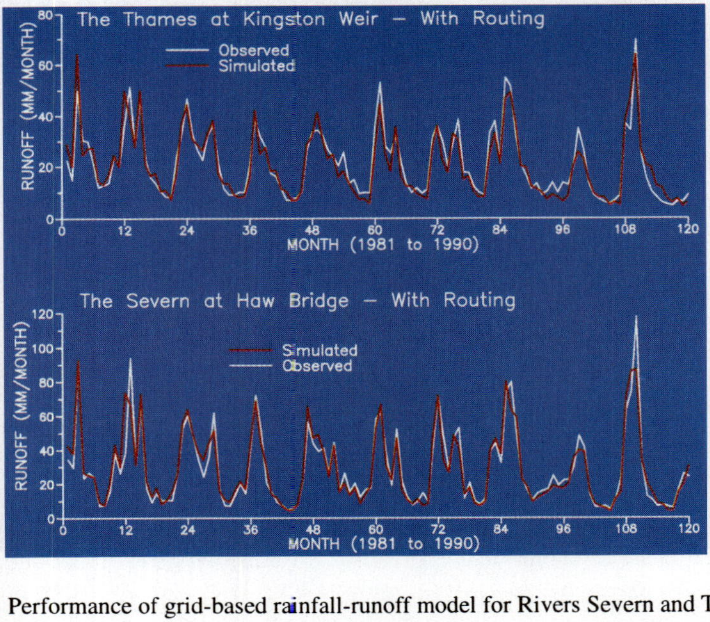

Fig. 21. Performance of grid-based rainfall-runoff model for Rivers Severn and Thames.

a hybrid approach of physics-based modelling of near-surface processes and metric analysis to define sub-surface routing. We need to extend this analysis to larger scales, but this methodological step has exciting potential for a range of hydrological modelling problems.

Conclusions

Science and sustainability

We have moved from the scale of a pore in a groundwater system to the global scale, and I have attempted to show that the contemporary problems of environmental management are placing increasing demands on hydrological science. However, significant progress has been, and is being, made.

Sustainable development is a concept that has received much recent political attention in international initiatives, such as the Rio Earth Summit and the associated Agenda 21. This idea has long been applied to water resource development, although more work is needed to set this concept in a broader perspective for management of the hydrological cycle. Our work in the arid zone, for example, aims to develop methods by which sustainable water resource development can be quantified in climatic regimes influenced by extreme temporal variability. And we have recently been evaluating the impacts of systematic change in rainfall patterns on flood frequency in an area of western Ireland (Wheater & Chandler, 1997). This promises to open a major area of investigation of non-stationarity effects in environmental management in general and, in particular, water resources.

The concept of sustainability needs to be extended further by hydrologists to include water quality and the management of wastes, to the interdependence between hydrology and ecology, and to the global environment. These are issues of current research. However, sustainability can only be achieved on the basis of scientific understanding. In tracing the development of scientific hydrology, we can point to notable successes and its increasing role in resolving practical problems. Continued scientific development is required to underpin effective management and resolve contemporary problems. This will require sustained effort and high quality research.

Crisis and conflict

The potential for conflict exists in the allocation of water resources, particularly where there is an international dimension. "Water wars" has become a well-used phrase which indicates the sensitivity of economies to water shortage and the potential international conflict concerning shared water resources. There are obvious sources of tension. For example, if we consider the Nile, 85% of the resource comes from Ethiopia, which has no rights to water under the Nile waters agreement. And the balance of allocation between upstream and downstream users of the Tigris and Euphrates is in the process of change, with unpredictable consequences. However, many workable international agreements exist, not least the recent accommodations between Israel and Jordan and India and Pakistan. It is the task of science to help produce sustainable management solutions; but it is the task of the international community to resolve conflicts.

More complex is the conflict between environment and development. This can occur on a local scale, such as our over-abstracted catchments in the UK, at regional scale (acid rain), or indeed, on the global scale (the greenhouse effect and climate change). The value we place on the environment is essentially a political issue and conflict can clearly arise where different societies take a different view. This is evident in disagreements between the developing and developed world. The recent dispute between Hungary and Slovakia over the Danube dams was engendered by the dramatic change of governments in eastern Europe, and the perception in Hungary that ecological concerns should receive greater recognition in the balance between development and environmental protection. Such issues are set to become an increasingly important aspect of international relations, including policy at the global level.

The responsibility of science is to provide appropriate tools for environmental management, to inform political judgement, and to warn of environmental damage and the consequences of unsustainable development. This requires integrity and impartiality and, on occasions, personal courage. There are careers that have been sacrificed in the former eastern Europe in voicing environmental concerns which were not consistent with state policy. Let us hope that such cases are in the past.

The recent (September 1997) judgement of the International Court of Justice on the Hungary/Slovakia dispute is likely to be an important international legal precedent in establishing water rights for international rivers, and in balancing economic development and environmental impact. Beyond the judgement itself, it is essential that an appropriate international forum commands the respect to arbitrate in complex international environmental issues — if future interstate conflicts are to be resolved amicably. In this context, too, the recent judgement is an important step.

In this review, we have considered the contemporary and important problems of water in a changing world. I hope that I have demonstrated the need to turn to science to achieve sustainability, as well as to provide the insight to resolve crisis and conflict at local, national and international levels. Hydrologists have a crucial role to play in the development and application of their science. We have made significant progress, but there are many important challenges ahead of us.

Acknowledgements

The development of hydrology at Imperial College commenced with the appointment of Peter Wolf in 1949. He pioneered the postgraduate programme, and it was taken forward by a number of distinguished staff, including Terence O'Donnell, Enda O'Connell and Mike Hall. I would like to express my appreciation to them for their legacy, and to my former colleagues, Elizabeth Shaw and Paul Jowitt. Their support in my early years at Imperial College was invaluable, as was the continuous encouragement of the late Roger Perry.

However, the content of this lecture is due primarily to a large number of dedicated and talented research students and research assistants, to whom space precludes full acknowledgement. They have provided enthusiasm, inspiration and much hard work, and have done a great deal to further the development of hydrology, both at Imperial College and in their subsequent careers. I am fortunate to have retained Adrian Butler and Christian Onof as academic colleagues, recently joined by Matthew Lees. It has also been a great privilege to work jointly with a number of extremely distinguished colleagues, including Bruce Beck, Valerie Isham, Sir David Cox, Ignatio Rodriguez-Iturbe and Gedeon Dagan. Their insight, scholarship and enthusiasm have been an important contribution to many aspects of my research.

References

Andersen, N.J. & Wheater, H.S. (1998) Sustainable development of alluvial groundwater in the sand rivers of Botswana. In *Proc. BHS Int. Conf. Catchment Management and Resource Assessment in Dry Areas, Exeter* (in press).

Bear, J. (1979) *Hydraulics of Groundwater*. New York: McGraw-Hill.

Beck, M.B., Kleissen, F.M. & Wheater, H.S. (1990) Identifying flow paths in models of surface water acidification. *Rev. of Geophys.*, **28**(2), 207–230.

Beven, K. & Binley, A.M. (1992) The future of distributed models: model calibration and perceived uncertainty. *Hydrol. Proc.*, **6**, 279–298.

British Hydrological Society (1994) *Sustainability in a Changing World: The Key Role of Hydrology*. London: BHS.

Burne, S., Butler, A.P. & Wheater, H.S. (July, 1997) Results of optimisation for WATBAL model. *ICON Report to BNFL.*

Burne, S., Wheater, H.S., Butler, A.P., Johnston, P.M., Wadey, P., Shaw, G. & Bell, J.N.B. (1994) Radionuclide transport above a near-surface water table. I — An automated lysimeter facility for near-surface contaminant transport studies. *J. Env. Quality*, **23**(6), 1318–1329.

Butler, A.P., Chen, J., Aguero, A., Edlund, O., Elert, M., Kirchner, G., Raskob, W. & Sheppard, M. (1998) Model validation studies of water flow and radionuclide transport in vegetated soils using lysimeter data. *J. Env. Radioactivity*, **42**(2–3), 271–288.

Butler, A.P. & Wheater, H.S. (1990) Model sensitivity studies of radionuclide uptake in cropped lysimeters. *Nirex Safety Series Report*. UK Nirex Ltd.

Calder, I.R. (1993) Hydrologic effects of land-use change. In Handbook of Hydrology, ed. D.R. Maidment, pp. 13.1–13.50. New York: McGraw-Hill.

Chapman, P.J., Reynolds, B., Wheater, H.S. (1993) Hydrochemical changes along stormflow pathways in a small headwater moorland catchment in mid-Wales. *J. Hydrol.*, **151**, 241–265.

Cox, D.R. & Isham, V.S (1980) *Point Processes.* London: Chapman and Hall.

Dagan, G. (1989) *Flow and Transport in Porous Formations.* Berlin: Springer-Verlag.

Gelhar, L.W. (1993) *Stochastic Sub-surface Hydrology*. New Jersey: Prentice-Hall.

Hutchins, M.G., Wheater, H.S. & Beck, M.B. (1995) The role of natural tracers in the understanding of catchment hydrochemistry within a conceptual hydrological framework. In *Tracer Technologies for Hydrological Systems. Proc. Boulder Symp., IAHS, Publn. No. 229,* pp. 267–276.

Huyakorn, P.S. & Pinder, G.F. (1983) *Computational Methods in Subsurface Flow*. New York: Academic Press.

Ibitt, R.P. & O' Donnell, T. (1971) Fitting methods for conceptual catchment models. In *Proc. Amer Soc. Civ. Engrs., 97 (HY9)*, pp. 1331–1342

Jolley, T.J. & Wheater, H.S. (1996) A large-scale grid-based hydrological model of the Severn and Thames catchments. *J. Water & Envtl Mgmt*, **10**(4), 253–262.

__________ (1997) The introduction of runoff routing into large-scale hydrological models. *J. Hydrological Processes* (in press).

Karavokyris, I., Butler, A.P. & Wheater, H.S. (1990) The development and validation of a coupled soil-plant-water model (SPWI). *Nirex Safety Series Report NSS/R225*. UK Nirex Ltd.

Kirkby, M.J. (ed., 1978) *Hillslope Hydrology*. New York: Wiley.

Kleissen, F.M., Beck, M.B. & Wheater, H.S. (1990) Identifiability of conceptual hydrochemical models. *Wat. Resour. Res.,* **26**(2), 2979–2992.

__________ (1997). A segmented optimisation strategy for parameter estimation for conceptual catchment models (submitted for publication)

Koide, S. & Wheater, H.S. (1992) Sub-surface flow simulation of a small plot at Loch Chon, Scotland. *J. Hydrol. Proc.*, **6**, 299–326.

Lees, M.J., Price, N., Wheater, H.S. & Peach, D. (1998) A flood simulation model for the Galway Bay region of Ireland. *Proc. Int. BHS Conf. Hydrology of Environmental Hazards, Exeter* (in press).

Mason, B.J. (1992) *Acid Rain — Its Causes and Effects on Inland Waters*. Oxford: Clarendon Press.

Monteith, J.L. (1965) Evaporation and the environment. *Symp. Soc. Expl. Biol.*, **19**, 205–234.

_________ (1973). *Principles of Environmental Physics*. London: Edward Arnold.

Muscutt, A.D., Reynolds, B. & Wheater, H.S. (1993) Sources and controls of aluminium in storm runoff from a headwater catchment in mid-Wales. *J. Hydrol.*, **142**, 409–425.

Natural Environmental Research Council (1975) *Flood Studies Report*. Five volumes. National Environment Research Council (NERC).

Northrop, P.J. (1996) *Modelling and Statistical Analysis of Spatial-Temporal Fields*. PhD thesis. Department of Statistical Science, University College of London.

Onof, C. & Wheater, H.S. (1993) Modelling of British rainfall using a random parameter Bartlett-Lewis rectangular pulse model. *J. Hydrol.*, **149**, 67–95.

_________ (1994) Improvements of the modelling of British rainfall using a modified random parameter Bartlett-Lewis rectangular pulse model. *J. Hydrol*, **157**, 177–195.

_________ (1996a) Analysis of the spatial coverage of British rainfall fields. *J. Hydrol.* **176**, 97–113.

_________ (1996b) Modelling of the time-series of spatial coverages of British rainfall fields. *J. Hydrol.*, **176**, 115–131.

Onof, C., Faulkner, D. & Wheater, H.S. (1996a) Design rainfall modelling in the Thames catchment. *Hydr. Sci. Jnl.*, **41**(5), 715–733.

Onof, C., Northrop, P., Wheater, H.S. & Isham, V. (1996b) Spatio-temporal storm structure and scaling property analysis for modeling. *J. Geophys. Res.*, **101**, 415–426.

Osborn, H.B., Renard, K.G. & Simanton, J.R. (1979) Dense networks to measure conventive rainfall in the southwestern United States. *Wat, Resour. Res.* **15**(6), 1701–1711.

Palmier, L.R. (1995) *Hydrological Aspects of Landfill Waste Management — A Case Study of the Drigg Low-level Radioactive Waste Repository.* PhD thesis. Imperial College.

Penman, H.L. (1948) Natural evaporation from open water, bare soil and grass. *Proc. R. Soc. Ser. A.* **193**, 120–145.

_________ (1949) The dependence of transpiration on weather and soil conditions. *Soil Sci.,* **1**, 74–89.

Rodriguez-Iturbe, I., Cox, D.R. & Isham, V.S. (1987) Some models of rainfall based on stochastic point processes. *Proc. R. Soc. London.*, **A410**, 269–288.

_________ (1988) A point process model for rainfall: further developments. *Proc. R. Soc. London.*, **A417**, 283–298.

Rutter, A.J., Kershaw, K.A., Robins, P.C. & Morton, A.J. (1971) A predictive model of rainfall interception in forests. I. Derivation of the model from observations in a plantation of Corsican pine. *Agric. Meteorol.*, **9**, 367–384.

Rutter, A.J., Morton, A.J. & Robins, P.C. (1975) A predictive model of rainfall interception in forests. II. Generalisation of the model and comparison with observations in some coniferous and hardwood stands. *J. Appl. Ecol.*, **12**, 367–380.

Samuel, C. R. (1999) Stochastic rainfall modelling of convective storms in Walnut Gulch, Arizona. Unpublished PhD Thesis, University of London.

Shuttleworth, W.J. (1993) Evaporation. In *Handbook of Hydrology*, ed. D.R. Maidment. New York: McGraw-Hill.

Tompkins, J.A., Gan, K.C., Wheater, H.S. & Hirano, F. (1994) Prediction of solute dispersion using a stochastic numerical methodology. *J. Hydrol.*, **159**, 105–123.

van Leeuwen, M., te Stroet, C.B.M., Butler, A.P. & Tompkins, J.A. (1998a) Stochastic determination of capture zones of groundwater wells. *Wat. Resourc. Res.* (in press).

_________ (1998b) Stochastic determination of the Wierden capture zones. *Groundwater* (in press).

Wadey, P., Shaw, G., Bell, J.N. & Minski, J. (1994) Radionuclide transport above a near-surface water table. II. Vertical distribution of gamma activities within soil profiles in relation to wheat rooting density and soil-to-plant transfers. *J. Entl. Quality*, **23**(6), 1330.

Wang, H.F. & Anderson, M.P. (1982) *Introduction to Groundwater Modelling*. San Francisco: Freeman.

Warrilow, D.A., Sangster, A.B. & Slingo, A. (1986) Modelling of land surface processes and their influence on European climate. *Dynamical Climatology Technical Note No. 38.* UK Met. Office, Bracknell.

Wheater, H.S. (1995) *Scientific Evaluation of the Gabcikovo-Nagymaros Barrage System and Variant C, Vol. 2 Counter-Memorial of the Republic of Hungary to the International Court of Justice:* Ch1. Introduction pp. 1–3; Ch3. Surface and Groundwater 3.4–3.6: Groundwater, Groundwater Quality, Bank-Filtered Water Supplies pp. 76–122; Ch5. Soils, Agriculture, Forestry, Fishery 5.1 soils, 5.2 Agriculture pp. 164–181.

Wheater, H.S. & Beck M.B. (1995) Modelling upland stream water quality: Process identification and prediction uncertainty. In *Solute Modelling in Catchment Systems*, ed. S. Trudgill, pp. 305–324. New York: Wiley.

Wheater, H.S. & Brown, R.P.C. (1989) Limitations of design hydrographs in arid areas — an illustration from south-west Saudi Arabia. *Proc. Second Nat. BHS Symp.*, 3.49–3.56.

Wheater, H.S. & Butler, A.P. (1989) Groundwater modelling of the Konkola mine. *Find Report to Zambian Consolidated Copper Mines.* Imperial College.

Wheater, H.S., Butler, A.P., Stewart, E.J. & Hamilton G.S. (1991) A multivariate spatial-temporal model of rainfall in S.W. Saudi Arabia. I. Data characteristics and model formulation. *J. Hydrol.*, **125**, 175–199.

Wheater, H.S. & Chandler, R.E (1997) Gort flooding study — rainfall aspects. *Imperial College Report to Southern Water Global.*

Wheater, H.S., Isham, V.S., Cox, D.R., Chandler, R.E., Kakou, A., Northrop, P.J., Oh, L., Onof, C. & Rodriguez-Iturbe, I. (1998a) Spatial-temporal rainfall fields: modelling and statistical aspects. *J. Hydrol. Earth Sys. Sci.*, special edition (to appear).

Wheater, H.S., Jakeman, A.J. & Beven, K.J. (1993b) Progress and directions in rainfall-runoff modelling. In *Modelling Change in Environmental Systems*, ed. A.J. Jakeman, M.B. Beck & M.J. McAleer, pp. 101–132, New York: Wiley.

Wheater, H.S., Jolley, T.J., Onof, C., Mackay, N. & Chandler, R.E. (1998) Aggregation, disaggregation and scale in hydrological modelling for atmospheric GCMs. *J. Hydrol. Earth Sys. Sci.*, special issue (to appear).

Wheater, H.S., Jolley, T.J. & Peach, D. (1995) A water resources simulation model for groundwater recharge studies: an application to Wadi Ghulaji, the Sultanate of Oman. *Proc. Int. Conf. Wat. Resourc. Management in Arid Countries, Muscat*, pp. 502–510.

Wheater, H.S., Langan, S.J., Brown, A. & Beck, M.B. (1991a) Hydrological response of the Allt a Mharcaidh catchment — inferences from experimental plots. *J. Hydrol.*, **123**, 163–199.

Wheater, H.S., Larentis, P. & Hamilton, G.S. (1989) Design rainfall characteristics for SW Saudi Arabia. *Proc. Instn. Civ. Engrs. Part 2*, pp. 517–538.

Wheater, H.S., Onof, C., Butler, A.C. & Hamilton, G.S. (1991c) A multivariate spatial-temporal model of rainfall in S.W. Saudi Arabia. II. Regional analysis and model validation. *J. Hydrol.*, **125**, 201–220.

Wheater, H.S., Shaw, T.L. & Rutherford, J.C. (1982) Storm runoff from small lowland catchments in South West England. *J. Hydrol.*, **55**, 321–337.

Wheater, H.S., Tuck, S., Ferrier, R.C., Jenkins, A., Kleissen, F.M., Walker, T.A.B. & Beck, M.B. (1993a) Hydrological flow paths at the Allt a' Mharcaidh catchment *J. Hydrol.* Proc., **7**, 359–371.

Wheater, H.S., Tompkins J.A., Van Leewan, M. & Butler, A.P. (1998a) Uncertainty in groundwater flow and transport modelling — a stochastic analysis of well protection zones. *Proc. Geocomputation* (in press).

Zhou, W., Wheater, H.S., Johnston, P.M. & Dawe, R. (1993) Visualisation of migration and entrapment of dense non-aqueous phase liquids in a fractured porous medium. *IAH Memoires, Vol XXIV, Part 2*, pp. 902–911.

Zhou, W.F. & Wheater, H.S. (1995) Effect of aperture variation on two-phase flow in fractures. In *Groundwater Quality: Remediation and Protection, Proc. Prague Conf., IAHS Publn. No. 225*, pp. 173–182.

CHAPTER 3

MAN, MINERALS AND ENVIRONMENT

A.J. MONHEMIUS*

The Material Ages of Man

The history of mankind, from its primitive beginnings to today's complex society, is inextricably linked to our use of minerals and metals. This evening, I want to take a very brief and rather eclectic look at our use of these materials through the ages and to examine, in particular, the environmental effects of the extraction and processing of minerals and metals. What I hope to show is that there is much yet to be learnt from the detailed study of historical mining and metallurgical sites; knowledge which can be used to help us to design and operate today's and tomorrow's mines and processing plants in a manner which will minimise their adverse impacts on our increasingly stressed and fragile environment.

Archaeologists, of course, have long recognised the crucial link between man's historical development and his use of materials and they have divided pre-history into a number of periods — each named after the dominant material of the time — as shown in Fig. 1.

Long before man became a farmer, the nomadic tribes who lived as hunter-gatherers used minerals to fashion tools and weapons which they needed to hunt animals and to butcher them for meat. The first

---

*Roy Wright Professor of Mineral and Environmental Engineering, T.H. Huxley School of Environment, Earth Science and Engineering, Royal School of Mines.

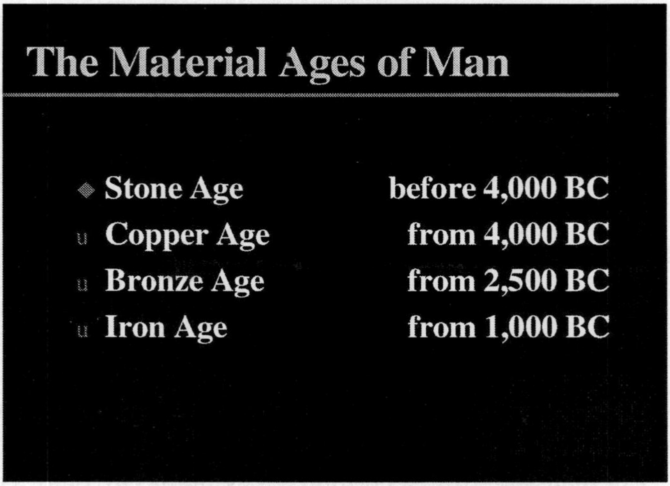

Fig. 1. The Material Ages of Man.

mineral products used by early man were those most readily found, namely stones lying around on the ground. The Stone Age lasted for many thousands of years, and during that long period man became highly skilled in the use of stone for making tools and weapons, and clay for making pottery and other utensils.

The only metal that came into use during the Stone Age was gold. The reason that gold was the first metal used by man is because it is one of only a handful of metals which exist in nature in the elemental state. Most of the metals that we are familiar with today are found as minerals, which are chemical compounds where the metal is combined with other elements, such as oxygen in metal oxides, or with sulphur in metal sulphides. Gold, however, is found as native gold, yellow and shiny.

As well as being quite rare, gold is also very soft and malleable. Compared with stone, it is a very poor material from which to fashion tools or weapons. Hence in Neolithic times, its use was confined to decorative purposes — mainly jewellery to show status and wealth.

The first material to seriously challenge the dominant role of stone in society was copper, and its alloys known as bronzes. There is some doubt as to when the Chalcolithic (Copper) and Bronze Ages began, but it was at least some 6,000 years ago that copper came into regular use in the more advanced civilisations of that time.

Knowing the chemical and geological behaviour of copper, it is not at all surprising that it was the first metal to be used for technological purposes. Like gold, copper is one of the few metals which is found in the elemental state. However, by no means is all copper found as metallic copper. In fact, as a proportion of the total amount of copper in the earth's crust — the vast majority of which exists as copper minerals — elemental copper is quite rare. However, where the geological conditions are right and elemental copper has been formed, it can be quite common locally and could have been found in sufficient quantities to be useful to early man. Also like gold, copper is quite malleable, which means that it can be beaten and shaped with simple tools into pots, cooking utensils, shields and weapons. However, because of its softness, a knife or sword made of copper would soon blunt and would be no match for a well-honed flint knife. The latter, being made of a very hard material, would keep its sharp cutting edges for much longer.

Eventually, these problems were overcome by the discovery that when copper was mixed with certain other metals — such as arsenic or tin — to form alloys called bronzes, the hardness of the resulting material was much greater. Hence the Copper Age became the Bronze Age.

The use of bronzes is a very significant point in the development of mankind because it marks the beginning of what today we generically call "technology". In order to produce bronzes, man had to master the art (later to become the science) of smelting — the use of high-temperature reactions to convert minerals into metals.

That copper and its alloys were the first metals to be produced by smelting is, again, not surprising. From a chemical point of view, copper oxides and copper sulphides are transformed into copper metal relatively easily. The chemical bonding between copper and oxygen and copper and sulphur is not very strong and the chemical conditions needed to break these bonds are achieved relatively easily. For example, heating copper oxide in a wood fire can lead to the formation of copper metal:

$$CuO_{(s)} + C_{(wood)} = Cu_{(s)} + CO_{(g)}\,.$$

Thus, the accidental discovery of this very important reaction is highly probable — almost inevitable, some would claim — given enough time.

Another reason why copper was the first metal to be produced by smelting is that many copper minerals are brightly coloured and highly visible. Thus, for the forbears of today's exploration geologists, finding these copper minerals was relatively easy. Once found, these minerals had to be extracted from the ground where they lay buried. Hence copper mining was born. The mines supplied the minerals to the smelters who converted the minerals into copper and bronze for the coppersmiths, who then made the utensils to supply the market, which rapidly grew as man discovered more and more ingenious ways of using these new miracle materials (Fig. 2). Thus was the beginning of the production

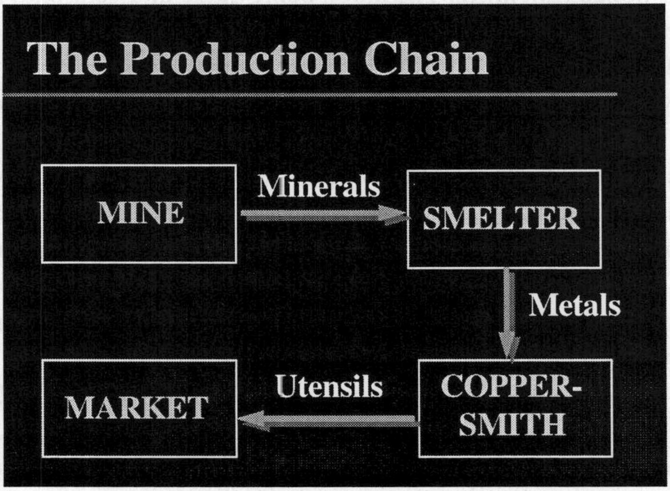

Fig. 2. The Production Chain.

chain from minerals in the ground through to metals in the service of man, with which today we are so familiar that we simply take it for granted.

Mining and Environment: The Early Years

The environmental effects of this production chain began to be manifest even in the Bronze Age. At first, copper production was limited because the mines and smelters were very small. Even so, they had environmental impacts, but generally these were very localised. However, in certain parts of the world, where civilisation was more advanced, the demand for metals was significant. For example, the copper mines at Timna on the Sinai Peninsular, which were the main source of copper for the Egyptian New Kingdom in the 14th to 12th centuries BC, covered an area of tens of square miles and there is evidence of thousands of ancient mine shafts.

By the middle of the first millennium BC, mining was a well-established industry in Greece and, indeed, the wealth of Athens during its golden age was largely derived from the silver mines at Laurium, which lie about 40 km south of the city of Athens on Cape Sounion. In 500 BC, there was a highly developed mining and smelting industry at Laurium and many of the ancient mine shafts and ore dressing installations are still in existence today. The ore was concentrated by a washing process, which is illustrated in Fig. 3, and the concentrate, which was silver-bearing lead sulphide, was then smelted in small blast furnaces (shown in Fig. 4) in a rather sophisticated, two-stage process which eventually yielded the silver to fill the treasuries of Athens. In Fig. 5, one of the furnace operators can be seen and it is evident that health and safety was an issue even in those times. The operator is wearing a hard hat, presumably to comply with the health and safety regulations in ancient Athens. The fact that he is otherwise stark naked seems not to have been a problem! The environmental effects of the ancient mining and smelting of lead at Laurium are not documented, but today the whole

Fig. 3. The ore concentration process in ancient Laurium, Greece (*Source: C.E. Conophagos, Le Laurium Antique; Courtesy: NTUA*).

Fig. 4. Smelting silver-bearing lead ore in ancient Laurium, Greece (*Source: C.E. Conophagos, Le Laurium Antique; Courtesy: NTUA*).

Fig. 5. Ancient Greek furnace operator (*Source: Berlin Museum*).

area is certainly heavily polluted. Most of this pollution, however, was caused by modern mining and smelting carried out in this area during the first half of this century.

Moving swiftly forward another two millennia, we arrive in Central Europe during the 16th century. Here, we find the first textbook on mining and metallurgy, as it was then practised in Saxony and the surrounding areas. The book is the famous *De Re Metallica* written by Georgius Agricola and published in Basel in 1556 (see Fig. 6). We have several early editions of this book here in the Imperial College Library in the Annan Collection of historical mining books, and the library has mounted a small exhibition of some of these to coincide with this lecture.

De Re Metallica was translated from the original Latin by Herbert Hoover, who was a mining engineer before he became President of the

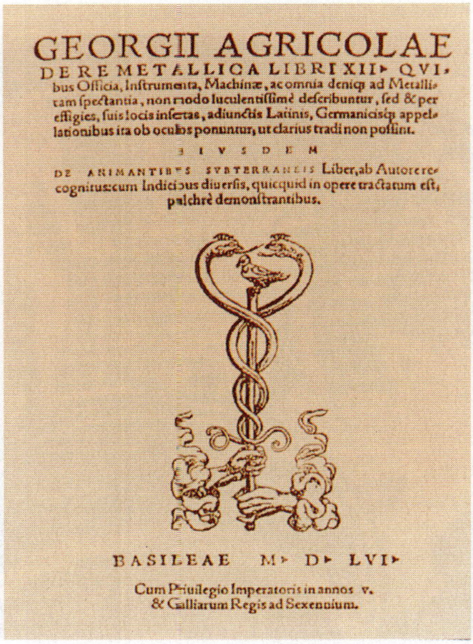

GEORGII AGRICOLAE
DE RE METALLICA LIBRI XII. QVIbus Officia, Instrumenta, Machinæ, ac omnia deniq; ad Metallicam spectantia, non modo luculentissimè describuntur, sed & per effigies, suis locis insertas, adiunctis Latinis, Germanicisq; appellationibus ita ob oculos ponuntur, ut clarius tradi non possint.
EIVSDEM
DE ANIMANTIBVS SVBTERRANEIS Liber, ab Autore recognitus: cum Indicibus diuersis, quicquid in opere tractatum est, pulchrè demonstrantibus.
BASILEAE M. D. LVI.
Cum Priuilegio Imperatoris in annos V. & Galliarum Regis ad Sexennium.

Fig. 6. Frontispiece of *De Re Metallica*, 1556.

United States. The translation runs to some 600 pages and contains over 250 original woodcuts. It is a complete and detailed account of the methods and processes used to mine and extract metals at that time, covering geology, mining, smelting and assaying practice.

It is divided into twelve parts or books, and revealingly, the whole of Book 1 is taken up with philosophical arguments and justifications as to why man should wish to use metals at all:

> "... there has always been the greatest disagreement amongst men concerning metals and mining, some praising, others utterly condemning them ..."

In the Middle Ages, there was a widespread school of thought that God, in his wisdom, had buried metals in the ground and there is where they should be left. According to that philosophy, metals were creations of the devil which caused nothing but trouble. Gold and silver led to jealousy and avarice. Copper and iron were made into weapons for killing and lead was used for bullets. Thus, the argument went that mankind would be much better off without metals and could manage very well without them.

Agricola, however, was not of this opinion and went to great lengths to list in his book all the beneficial uses for metals. He was, nevertheless, very much aware of the environmental effects of mining at that time:

> " … the strongest argument of the detractors is that the fields are devastated by mining operations. Also, they argue that the woods and groves are cut down, for there is need of an endless amount of wood for timbers, machines, and the smelting of metals. And when the woods and groves are felled, then are exterminated the beasts and birds, very many of which furnish a pleasant and agreeable food for man. Further, when the ores are washed, the water which has been used poisons the brooks and streams, and either destroys the fish or drives them away … "

Environment and Modern Mining

Thus even in the Middle Ages, mining had its vociferous detractors and, as we all know, the same situation still exists today. What has dramatically changed over the centuries, however, is the size and scale of mines. Those of the 16th century were quite small, being largely dug by hand with the help of only horse power and water power. Since the introduction of mechanically powered machinery into mining, first steam engines in the 18th and 19th centuries, followed by internal combustion engines and electric motors during this century, the machinery used in mining has become ever larger and more powerful. The mines themselves have also increased in size until today, our largest mines have grown to monstrous proportions.

The biggest man-made hole in the ground is claimed to be the Bingham Canyon copper mine in Utah (Fig. 7). This lies not many miles from

Fig. 7. The Bingham Canyon Mine (*Courtesy: Rio Tinto plc.*).

the Black Rock Desert where the British car "SSC Thrust" recently broke the sound barrier. Operations began at Bingham Canyon at the beginning of this century and since then, over five billion tons of rock have been moved, creating a hole two and a half miles across and half a mile deep, which is said to be visible by the naked eye from the orbiting space shuttles.

While Bingham Canyon is an extreme example of the size of modern mines, it nevertheless graphically illustrates the increase in the scale of mining operations which has occurred, particularly in the second half of this century. This has happened so that the industry can continue to supply the burgeoning demand for metals, while at the same time keeping costs in check by making economies of scale.

The environmental impact of mines has grown with their size. This has been matched by the growth in the bureaucracy involved in obtaining mining permits. For major projects these days, the planning and environmental permitting procedures can take several years. Some of the main issues that have to be addressed during the planning stages of new mines include land ownership, site access, dumping of waste rock, disposal of tailings, generation of noise, vibrations from blasting, generation of dust, potential for pollution of surface and ground waters. It is interesting to note that over 400 years ago, Agricola gave very similar advice to prospective miners:

> "Now a miner, before he begins to mine the veins, must consider seven things, namely, the situation, the conditions, the water, the roads, the climate, the right of ownership, and the neighbours".

Acid Mine Drainage

One of the most difficult environmental problems arising from mining operations is the phenomenon known as acid mine drainage. This occurs when sulphide minerals, especially the iron sulphide known as pyrite, are exposed to the atmosphere by mining operations. Over a period of time, a natural process of oxidation driven by bacterial action takes place which causes pyrite to break down into sulphuric acid and dissolved iron. The resulting acidic iron-bearing water is called acid mine drainage (AMD) and this is capable of creating considerable environmental damage.

An extreme example of destruction caused by AMD can be seen at an abandoned copper/cobalt mine called Kam Kotia in northern Ontario. Here acidic drainage from tailings or waste left after the mining and concentration process has devastated the surrounding forest, as illustrated in Fig. 8. This mine was operated during World War II and it was closed down at the end of the 1940s. Hence the acidic waters have been produced at Kam Kotia for the last 50 years and they are likely to continue for many more decades and even centuries. They will, in fact, continue until all the pyrite in the tailings has eventually been fully oxidised. This illustrates one of the most pernicious aspects of AMD — its longevity.

Pyrite is a very common mineral of widespread occurrence. It is found not only in metal mines, but also in most coal mines. Thus, acid mine drainage is a problem faced by many mines around the world, except

Fig. 8. Abandened mine site at Kam Kotia, Ontario, Canada.

those in very arid climates. During the working life of a mine, acidic drainage is usually kept under control by water treatment processes which are a normal part of the mine's day-to-day operations. The adverse effect on the environment is often not manifested until after the end of the life of the mine, when all the reserves of useful ore have been exhausted and the mine has been closed down. Like most holes dug in the ground, mines have a tendency to fill with water. When a mine is operating, it is kept dry by continuous pumping to remove water from the working areas. When a mine is abandoned, the pumps are switched off and removed and the mine will slowly fill with water. In many cases, because of the formation of acid mine drainage, the water in the mine will become acidic. Eventually, it may begin to flow into local water courses, resulting in environmental damage.

A graphic example of this occurred in this country in Cornwall, just over five years ago. The story began in early 1991 when the owners of the Wheal Jane Mine, one of the last operating tin mines left in Cornwall, gave up the struggle of trying to keep the mine operating profitably against a background of severely depressed tin prices, and finally switched off the de-watering pumps. Throughout 1991, the Wheal Jane mine gradually filled with water until suddenly, in January 1992, the water pressure in the mine caused a plug in a long-forgotten tunnel to give way. This allowed millions of gallons of acidic water to flow out of the mine overnight, into the Carnon River, and hence into the Carrick Roads outside Falmouth and into Falmouth Bay itself. When the iron-bearing acidic mine water mixed with the sea water, the iron immediately precipitated as ferric hydroxide, a gelatinous brown material which turned the sea turbid (Fig. 9).

Fig. 9. Effect of acid mine drainage in the sea near Falmouth, England (*Courtesy: Environment Agency and Knight Piésold Ltd.*).

The Response to Environmental Pressures

So what, in this day and age, can be done about mining and its effects on the environment? We hear much talk of sustainable development, recycling, clean technology, zero waste emissions and such like. We need, however, to understand some basic facts. Mankind needs its metals and minerals. The developing world, with its expanding populations and expectations of achieving a standard of living comparable with that enjoyed by the developed world, will have an insatiable demand for metals and minerals for the foreseeable future. This is likely to mirror or exceed the rate of growth seen in the developed countries in recent decades, as shown in Fig. 10. In the developed world, the increased use of recycling may moderate the rate of growth in demand, but undoubtedly more and more mines will need to be brought into production to satisfy the overall growth in market demand.

This has to be achieved in the light of the huge change in society's attitude to environmental issues that has taken place world-wide over

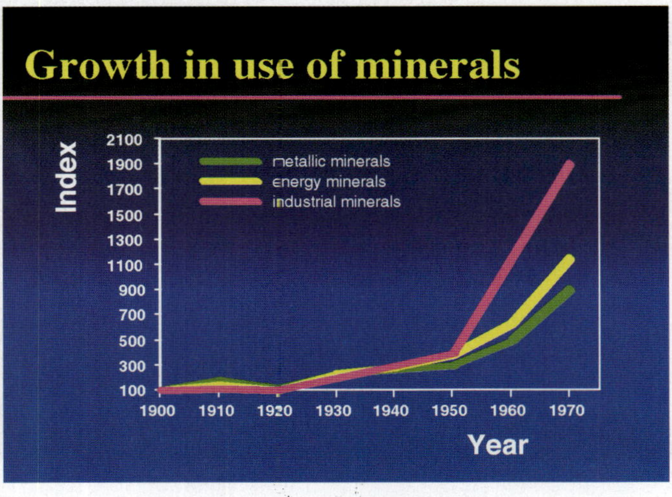

Fig. 10. Growth in demand for mineral products.

the last generation. Mining practices which were acceptable 25 years ago may no longer be tolerated today — ever higher standards are demanded, and rightly so. The mining industry today, however, suffers from a legacy of sins in the past. Its current environmental performance is vastly improved over that of a generation ago.

The mining and metallurgical industry was one of the first targets of the fledgling environmental movement when it began in the early 1960s. Hence the industry has had 30 years of ever-increasing pressure — legislative, regulatory and societal — to clean up its act. The enormous growth in the total number of environmental laws introduced in the United States during the 1970s and 1980s, since the Environmental Protection Agency was established, is shown in Fig. 11 The result of all these pressures is that all the major international mining companies now give environmental matters the highest priority, usually with a main board director having overall responsibility for the environmental performances of the company's mines around the world.

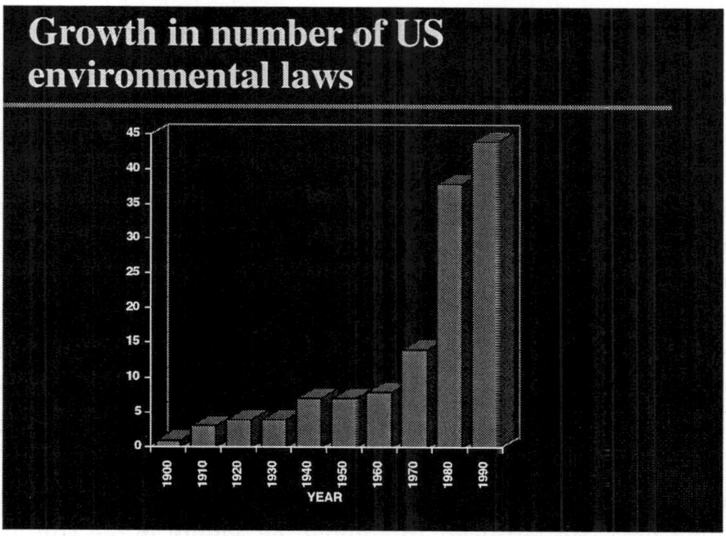

Fig. 11. Growth in the total number of US environmental laws.

In spite of the vast improvements that have taken place, it cannot be claimed that the industry no longer has an image problem. It is in the nature of the business that small companies with little capital are regularly being started up — the so-called junior mining companies. Some of these, by no means all, tend to cut corners in order to save scarce capital, sometimes with dire environmental consequences. And, of course, even in the best resourced operations, accidents can happen and these can also have undesired and unfortunate effects on the environment and generate bad publicity.

Restoration of Sites

One of the main problems facing the industry as a whole is the legacy of closed mining operations and the continuing environmental impact they have on their local areas. The consequences of closing down a mine when all the ore has been extracted depends very much on the type of mine, whether it is underground or surface, for example, and the way in which it has been operated and managed throughout its working life.

In some older mining areas, major restoration programmes are under way. One of the most ambitious is at Sudbury in Ontario, where nickel mining has been carried out since the begining of the century. The landscape around Sudbury used to be famous for the fact that the vegetation had been destroyed for miles around, due to a combination of over-extraction of timber, plus destruction by sulphur dioxide from the many nickel smelters in the area. In fact, the early astronauts did part of their training for the lunar landings in the Sudbury area, since this was the closest resemblance to the surface of the moon here on earth. About ten years ago, an ambitious programme to re-green the Sudbury area was begun. This is being undertaken by the major metal companies in the area, such as INCO and Falconbridge, together with community action by schools and other local groups. With a combination of soil conditioning by adding lime and fertilisers to combat the acidity in the soil, together with a huge programme of planting millions of saplings, the whole area

Fig. 12. Restoration of contaminated land at Sudbury, Ontario, Canada. The effect of soil conditioning and replanting can be seen in the foreground.

is gradually being transformed back to something like its original state (see Fig. 12).

In the case of large open-pit mines and quarries, restoration of the site, that is, returning it to its original pre-mining state, is clearly impossible and other solutions have to be sought. In the case of the huge Bingham Canyon mine, which we saw earlier, the Americans have been very clever. They have declared the mine to be a national monument, although it is still a fully operational mine. Hence there is now no question of restoring the site, which has become a popular tourist attraction.

The Island Copper mine, which is another large open-pit mine on the northern tip of Vancouver Island, used a rather unique solution for closure. This mine, which began operations in the late 1960s, ran out of ore last year after 25 years of operation. The mine was situated very close to the shoreline of Rupert Sound, a fjord on the west coast of Vancouver Island. From the beginning, this was a controversial project which underwent what, at the time, was the most exhaustive public planning inquiry ever held in Canada before it was given the mining permit to proceed. Incidentally, the principal opponent of the mining company at the planning inquiry was a young PhD student at the University of British Columbia, Patrick Moore, who subsequently went on to found and to become the first executive director of Greenpeace.

The most controversial aspects of the Island Copper project were the proposals for the disposal of waste rock and tailings from the mine. The plan was to dispose of both of these into Rupert Sound. The waste rock was to be used to reclaim land along the shoreline, while the tailings were to be disposed of some 200 yards offshore onto the deep bottom of the Sound. These proposals were eventually agreed upon, subject to stringent environmental monitoring procedures.

Mining, then, proceeded for a quarter of a century with minimal adverse effects on the ocean ecology until the mine ran out of ore and was finally closed in July of 1998. The closure plan for the mine called for it to be flooded with sea water from the Sound. This was done in spectacular fashion by cutting a channel from the Sound to the mine and creating an artificial waterfall which flowed for over 30 days until the mine was filled with sea water (see Fig. 13).

Research in the United Kingdom

Coming closer to home to look at the situation in the United Kindom (UK), we do not have any of these very large open-pit metal mines in this country, although some of our quarries are of a significant size. What we do have, however, is the legacy of a metal mining industry — which was the largest in the world in the middle of the last century — and a coal mining industry. The latter was also large by world standards as recently as the 1980s. Since then, it has shrunk dramatically. Today, we have less than 20 coal mines still operating. Hence we have a landscape which, in certain parts of the UK, shows very obvious signs of past mining activities. We also have problems with some rivers being contaminated by acid mine drainage. This, as I explained earlier, is a

Fig. 13. Flooding of the Island Copper Mine, Vancouver Island, Canada. (a) During flooding, (b) after flooding (*Courtesy: BHP Canada Ltd.*).

long-lasting problem and one which, if anything, is getting worse due to the recent closure of so many coal mines around the country.

Happily, we are making some progress towards the development of effective long-term treatments for acid mine drainage. In particular, as a result of the Wheal Jane incident, this country is at present a world leader in the development of the technology known as passive treatment. This is the use of natural processes which occur in marshes and other wetland areas. It leads to the cleansing of water by the removal of dissolved metals and other pollutants by bacteria in the wetlands. This type of biological solution to the problem can be utilised in so-called engineered wetlands, which are designed to make maximum use of these biological processes for metal removal and the clean-up of acid mine drainage.

The world's most advanced passive treatment pilot plant was built at the Wheal Jane mine after the pollution incident at a cost to the British taxpayer of over one million pounds (see Fig. 14). For the past two-and-a-half years, this plant has been used to test the effectiveness of the concepts for the treatment of the Wheal Jane mine water. Although

Fig. 14. The Wheal Jane Constructed Wetlands Pilot Plant, Cornwall, England (Courtesy: Environment Agency and Knight Piésold Ltd).

this work has shown that the concept does work and the cocktail of polluting metals can be removed from the mine water by the wetlands, with minimal intervention from man, the surface area of land required per unit volume of water treated is quite large. In the case of Wheal Jane, the flow rate of water from the mine is so large that a full-scale wetland, big enough to treat the whole flow, is not practical, given the limited land area available in the Carnon River Valley.

The first phase of development work on this pilot plant will end shortly. It is then hoped that the plant will be handed over to a consortium of universities, of which Imperial College is one, to continue research to gain a fuller understanding of the biological processes at work in these wetlands and, through greater understanding, to improve their efficiency of operation and to widen their potential areas of application to different types of mine waters and other effluents.

Another topic of major concern for both operating and closed mines is the safe disposal of solid wastes arising from mining and processing operations. I have already discussed the long-term problems of acidic drainage which can originate from sulphide-bearing waste rock and processed tailings if these are not properly stored on the surface. Of even greater concern are solid residues which contain toxic elements such as mercury, cadmium, lead or arsenic, to name but a few. Over the past five or six years, my research group has been actively engaged in tackling this type of problem, in particular with an emphasis on the ways of disposing arsenic in an environmentally safe manner.

We have carried out an extensive study of the phases that can be precipitated from aqueous solutions containing mixtures of iron, calcium, arsenate and sulphate, these being the most common constituents of arsenic-containing process solutions. We have shown that by using hydrothermal precipitation, namely, temperatures between 150ºC and 200°C, a synthetic analogue of a common arsenate mineral can be produced. This mineral is scorodite, hydrated ferric arsenate ($FeAsO_4.2H_2O$), which is formed geologically by the weathering of arsenic-containing sulphide minerals.

We have tested our synthetic scorodite extensively in the laboratory and found it to be stable and practically insoluble. However, our laboratory tests are relatively short-term, even though they may last for several years, when we are looking for a material suitable for permanent disposal. Ideally we wish to be able to predict the stability of a material over the next 100 to 1000 years, and it is just not possible to do this with any confidence on the basis of short term laboratory tests. What is needed are examples of mining and other processed wastes that have been left exposed to the elements for various lengths of time, which we can study to build up a picture of how they behave over extended time periods. To do this, we can take advantage of our mining heritage and its long history, with which I opened this talk. By studying mining sites from different periods of history, from the Bronze Age to the present day, by carrying out careful sampling and detailed mineralogical and chemical examinations, it should be possible to establish which are the most stable compounds of the various metals that we wish to dispose of. Thus, we can use information gleaned from ancient mining practices to further help improve the environmental performance of our modern mining and processing operations.

Environmental Engineering

Finally, I would like to say a few words about Imperial College's response in terms of its teaching function to the changes brought about in the minerals industry by environmental pressures. The Royal School of Mines, which is one of the three colleges amalgamated at the turn of the century to form Imperial College, was itself founded in 1851. Thus, this college has been training people for the minerals industry — geologists, mining engineers and metallurgists — for nearly 150 years. The RSM is world famous and its graduates can be found in virtually every mining district around the globe, many of them holding key positions in the industry.

The minerals industry is unique in that it is always in a state of flux. All mines eventually run out of ore and have to be replaced by new mines working new ore deposits. Thus, in addition to established mines and processing plants, there are always new projects underway to find and develop these new ore deposits. Equally, there are always mines reaching the end of their working lives which are in the process of closing down. Today, environmental considerations impinge strongly on all stages of this cycle.

Some five or six years ago, we came to the conclusion that the time had come to add a fourth discipline to the three traditionally taught at the RSM — namely, geology, mining engineering and metallurgy — and so we started a new degree in environmental engineering — or to give it its full title — Environmental and Earth Resources Engineering. Earlier this summer, we graduated our first cohort of environmental engineers and we hope that they and their successors will take their place in the industry to help it to continue to supply the world's metals and minerals.

The College was greatly aided in the development of this new course by a generous donation from the Rio Tinto Corporation to endow a chair in environmental engineering applied to the minerals industry, of which I am proud to be the first incumbent. In making this endowment, Rio Tinto stipulated that the chair should be named in honour of Roy Wright, who was one of the three founders of the company back in the 1950s. In the beginning, RTZ was a junior company whose only asset was the Rio Tinto mine in southern Spain, a famous mine whose origins go back to Chalcolithic times. Roy Wright, together with the two other founder members of the company, Val Duncan and Mark Turner, steered RTZ on a path of spectacular growth which continued after his retirement. Today, Rio Tinto, as it is now known, is the world's biggest mining company.

Roy Wright was very enthusiastic about the new endowment at Imperial College that bore his name, and we were fortunate that he visited us in the Royal School of Mines several times in the latter years of his life to find out how things were progressing. Sadly, Roy passed away two years ago, but I am very pleased that he is represented on this occasion by his daughter, who is here with us in the audience tonight.

In conclusion, I hope that this evening I have been able to give you a flavour of the ways in which the ancient and vital industry of mining, which throughout its long history has led mankind in the development and application of new technologies, is responding to the desire of modern society to protect and enhance the global environment. I have also briefly outlined the ways in which Imperial College, with its long association with the minerals industry world-wide through the RSM, is adapting its teaching to produce engineers who are environmentally aware and have the knowledge and skills to tackle problems from the past, to operate today's mines and processing plants, and to design those of the future, in a manner which minimises their impacts on the environment, while they continue to produce metals and minerals in the quantities and qualities which are such a fundamental necessity for man's continuing existence and his further development of living standards on this earth.

CHAPTER 4

ROCK, SOIL AND HEALTH: MAN'S INTERFERENCE WITH THE GEOCHEMICAL ENVIRONMENT

IAIN THORNTON*

Perhaps it was the publication of the Bruntland Report in 1987 that made the first major impression on politicians, scientists and administrators that population pressure and economic development were making a major deleterious impact on the environment in which we live. Of course, the environment now takes its rightful place at the top of the international political and scientific agenda. Our own Prime Minister at the time, Margaret Thatcher, in her much publicised speech to The Royal Society on 27 September 1988, set the scene and provided the momentum for what has proven to be an explosion in political, scientific and public interest in the United Kingdom (UK).

We only need to open a newspaper to be reminded that the environment is part of our daily lives, green issues are the name of the game — green soap powders are on the shelves of our supermarkets and green unit trusts are prepared to invest our savings in an environmentally conscious manner.

This inaugural lecture has given me the opportunity to re-examine and assess the multi-faceted and multi-disciplinary research with which I have been concerned at Imperial College for the past 26 years. To take stock — and I plan to share with you some of the more exciting

*Professor of Environmental Geochemistry

aspects of that work, to share some of the broader issues it has raised and some of the many questions that remain unanswered.

But I am not here to talk about green issues, greenhouse gases or ozone depletion — but about another and less publicised part of the environment, very much at the interface of the geosphere and the biosphere. My area of interest over the past 30 or so years has, and continues to be, environmental geochemistry. This is concerned with complex interactions in the following systems:

1. Geosphere-hydrosphere-atmosphere-biosphere.
2. Rock/soil-water-air-life.

The rapidly advancing research area of geochemistry and health, which I will be discussing today, is primarily concerned with the recognition and understanding of relations between the composition of rocks, soils, dusts, air-borne materials, ground and surface waters, and the health of plants, animals and man.

However, natural processes — natural geochemical cycles — have been grossly disturbed by man and his activities. The world's population has risen from 1,000 million in 1800 to 5,000 million today and is expected to double again by the year 2100. Large-scale urbanisation and industrialisation has, and continues to, make massive changes — often deleterious to soil, water and biological resources — and to the quality of the human environment.

What exactly is this environment we talk about? I am sure we all perceive it in different ways. The shorter version of the Oxford English Dictionary defines the word environment as "the conditions or influences under which any person or thing lives is developed", but in reality, even the terrestrial environment covers a vast range of possibilities. For example, the white arabian oryx was recently re-introduced into the wild with some of the Sultan of Oman's personal funds within a 7,000 km^2 of the Jiddat Al Harasis desert in Oman. It is a flat limestone plateau with no roads, rivers, only a few millimetres of rain every two to three years and very little vegetation — rather like a huge dry parking lot. At the other extreme, the Aberdares National Park in Kenya has a high rainfall and rich vegetation and supports a diverse and healthy wildlife population, including lion, elephant and buffalo.

Man may, then, interfere and farm the land with crops and livestock, but this again encompasses a huge range of possibilities in relation to the climate, the soil and the culture of local populations. Of course, man has built towns and cities and we have created a very different kind of environment where people, traffic and buildings interact in different ways.

There are also the people who must both perceive and interact with their contrasting environments. They range from the Bedouin tribesman in the desert, minority communities living in remote villages in China, African villagers practising slash-and-burn agriculture in the Gambia, and the huge leap from these simple lifestyles to the so-called Western society.

We can see that some people still live close to the soil, from which those of us living in developed industrial countries are far removed. We can see the world today as a place of extreme diversities — those places where a simple rural existence allows geochemical cycles to operate in a relatively undisturbed fashion — those situations where the impact of man, certainly in our Western society, has almost completely overridden any natural impact of geochemistry on our lives. This reflects the difference between the developing and the developed world — man living close to the earth compared with man enjoying various levels of so-called sophistication.

My own entry into the study of this complex and diverse arena commenced when I went to the Gambia in West Africa as a young graduate soil scientist, to a country of simple agricultural practice then dependent on one cash crop — *arachis hypogea* — the groundnut. There, I was soon to discover that natural geochemical cycles would not adequately support successful agricultural production under a system of agriculture already under population pressure. However, a little phosphatic fertiliser applied to the groundnut crop would increase the yield by a factor of two or more, thereby ensuring financial sufficiency both for the farmer and for the country — a case of where man's interference with things natural was highly beneficial.

Then, in 1964, I came to Imperial College to join Professor John Webb, a man of intense energy and considerable vision — an excellent leader and, in due course, a close friend. This was the beginning of an exciting and personally very rewarding quarter of a century.

John and his team in the Applied Geochemistry Research Group were amongst the pioneers in the field of geochemical exploration. Research, mainly overseas, had shown that the combination of systematic sampling strategies and multi-element analytical techniques could be used to compile geochemical reconnaissance maps. Indeed, the concept of a multi-purpose geochemical atlas was John's brainchild — maps not only of use in finding mineral deposits, but of use to the agriculturist concerned with mineral deficiencies or excesses and to the epidemiologist concerned with nutritional and toxicological disorders in the human.

My brief was to study the applications of geochemical reconnaissance mapping to agriculture. With funding from NERC for exploratory studies into biogeochemistry and a territory defined by the Institute of Geological Sciences (now the BGS) covering 3,000 square miles in Devon and Cornwall, 3,000 in Derbyshire and 3,000 in North Wales for the production of trial geochemical maps, my role was to look for and interpret possible relationships between geochemical maps which showed the regional distribution of a range of elements, including some nutritionally essential ones such as Cu, Co, Mn and Zn, and some potentially toxic ones such as As, Cd and Pb, and the production and health of farm crops and livestock. Six of my first 12 months were spent on fieldwork, talking to veterinary surgeons, agricultural advisors, soil scientists and nutrition chemists. The farmers themselves were visited when deficiency or toxicity problems were recognised or suspected. Many thousands of soil and herbage samples were taken and, on some occasions, even faecal samples from the cows.

Geochemical Mapping

Following the production of an exploratory geochemical atlas of Northern Ireland — comprising a multi-element database for 9,000 stream sediment samples (Webb *et al.*, 1973) — funding was obtained from the Wolfson Foundation in 1968 to compile a geochemical atlas of England and Wales. My first role in this was to arrange the collection of samples and then safe transportation to the laboratories at Imperial College. Then I took over as co-ordinator of the project, with a team of 56 specially recruited students and four graduate geologists. In 28 hired mini-vans, this team covered 60,000 square miles of England and Wales in 10 weeks, collecting 50,000 samples of stream sediment at stream-road intersections. Multi-element analysis was undertaken over a period of 18 months under the direction of Michael Thompson and data processing by Richard Howarth. This led, in due course, to the publication of the *Wolfson Geochemical Atlas of England and Wales* — a prototype atlas — the first to be published in the world and the forerunner of many since (Webb *et al.*, 1978).

An example from this atlas is shown in Fig. 1 for the element copper. Patterns of blue colour represent low concentrations and delineate areas in which deficiencies in cereal crops occur. Patterns of high concentration are shown in deep red, reflecting an area of copper, tin and arsenic mineralisation in southwest England, where several hundred years of mining operations have led to widespread contamination of the surface environment.

Since the delivery of this lecture, multi-purpose geochemical mapping has continued to flourish. The prototype *Wolfson Geochemical Atlas of England and Wales* was the forerunner of further geochemical atlases compiled by the British Geological Survey (Plant & Thornton, 1986) in the UK and the many geochemical atlases that continue to be published elsewhere. Geochemical maps of Western Europe (Demetriades *et al.*, 1990), and indeed, a primary Global Geochemical Reference Database under IGBP Project 259 based on wide-spaced sampling over the entire land surface (Darnley *et al.*, 1995), are being planned.

Geochemical maps and atlases provide a useful time-dependent catalogue of baseline data showing the "natural distribution" of chemical elements, including metals relating to the composition of rocks and soils, and patterns of metal enhancement relating to mineralisation. Such information is of great help to those assessing the degree and extent of environmental

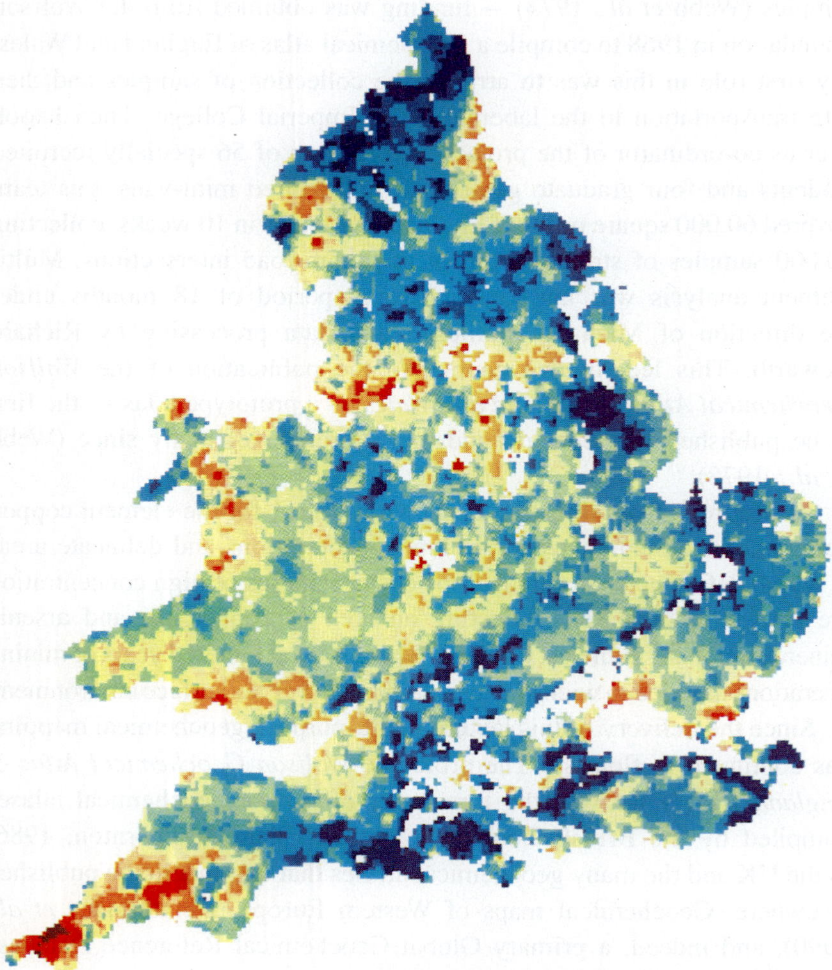

Fig. 1. Map showing the distribution of copper in stream sediments in England and Wales (Webb *et al.*, 1978).

pollution arising from human activities. They also provide primary reconnaissance data indicating areas in which nutritional deficiencies are likely to be found in agricultural crops, livestock and wildlife (Thornton, 1983).

Geochemical maps have proved invaluable in developing the discipline of geochemistry and health in Britain. Much of our research in environmental geochemistry has been concerned with the interpretation and application of geochemical maps — focusing postgraduate research of many PhD students into priority areas defined by these maps to study the sources, geochemical pathways in the soil-plant-animal system, as well as the effects of nutrient deficiencies and toxic metal excesses in the environment on plant, animal and human health.

A discussion meeting was organised at the Royal Society in 1978 by Professors Stanley Bowie and John Webb to communicate and evaluate the status of international research in this area and to identify future priorities (Bowie & Webb, 1980).

Applications to agriculture

Early work in this field is best illustrated by the interpretation of the geochemical map for molybdenum (Fig. 2). Molybdenum excess in soils may lead through the food chain to dietary excess in grazing animals. This, in turn, may cause molybdenosis or "teartness" in cattle with severe scouring or loss of condition (Ferguson *et al.*, 1943); moderate excess in the diet may cause depletion of Cu reserves in the ruminant and Mo-induced Cu deficiency, with symptoms of scouring, loss or change of coat colour, stunted growth, immaturity and unthriftiness.

This conditioned Cu deficiency is caused by the antagonistic action of Mo on the usage of dietary Cu. "Teartness" was first recognised in Britain on calcareous soils derived from the Lower Lias formation in Somerset. The source of the Mo was found to be a zone of interbedded black shales and limestone (Lewis, 1943). At about the same time, a similar condition was reported in parts of California on alluvial soils downstream of Mo mineralisation (Britton & Goss. 1946). In England

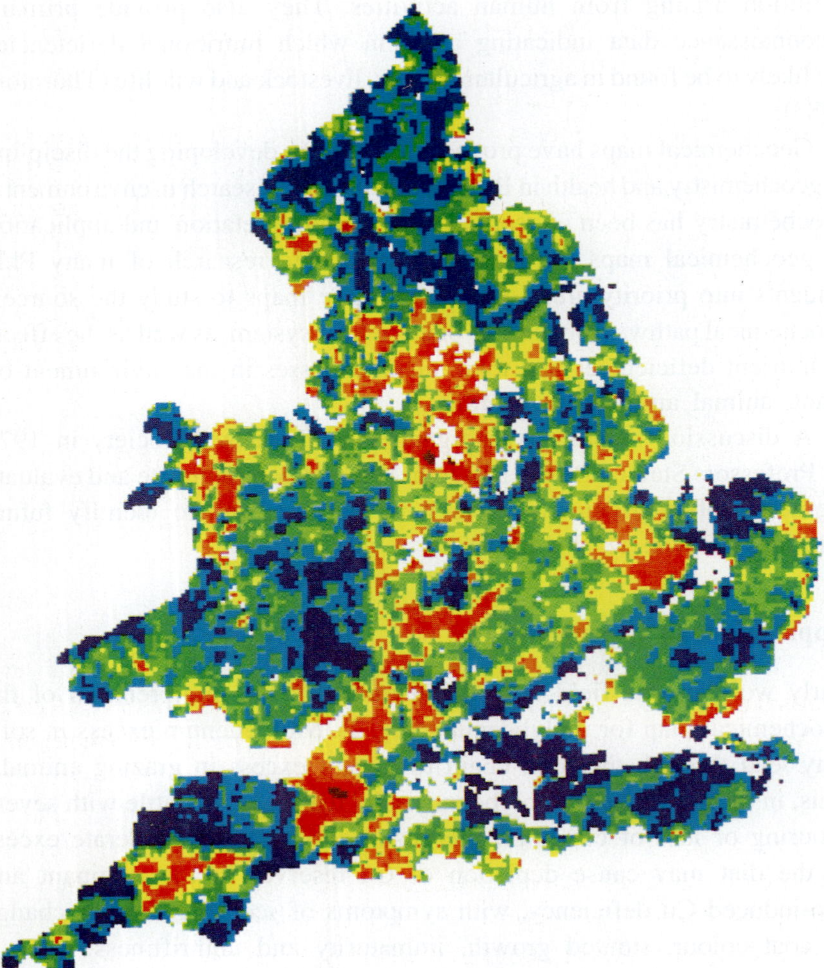

Fig. 2. Map showing the distribution of molybdenum in stream sediments in England and Wales (Webb *et al.*, 1978).

and Wales, Mo-induced Cu deficiency is often found when pastures contain 2 ppm Mo (D.M) or more (Thornton, 1977). It has also been experimentally shown that at this dietary level, Mo restricts the retention and storage of Cu in the liver (Suttle, 1981).

In Derbyshire, a county in central England, the molybdenum anomalies shown in Fig. 2 correspond with areas underlain by black Carboniferous shale and extend over far larger areas than those previously associated with clinical bovine hypocuprosis. Blood sampling programmes, covering 350 cattle from 26 herds, showed that 77% of the animals not fed minerals within the Mo anomalies were copper-deficient (Thornton *et al.*, 1972). Supplementation trials with injected copper showed responses in live weight gain in young cattle ranging from 10% to 70%, representing mean herd increases due to copper of 30–70 lbs per animal over a six-month grazing season (Thornton *et al.*, 1972). This provides a clear example of sub-clinical deficiency that will respond to treatment — which is economically important — and the extent of which has been identified as a result of geochemical mapping.

This geochemical reconnaissance map for England and Wales highlights the considerable extent of suspect high molybdenum land, which is now thought to exceed one million acres. Following the compilation of these national maps, the Agricultural Development and Advisory Service (ADAS) of MAFF set up Geochemical Liaison Committees in the (then) nine administrative regions of England and Wales to study and recommend effective methods of utilising and following up the geochemical information. As seen from Fig. 2, with the exception of southeast England, there are few regions in which molybdenum anomalies do not occur. Geochemical follow-up programmes were initiated in six of these "suspect" areas to check the validity of the data and, where necessary, to increase stream sediment sampling density with a view to presenting reconnaissance data for relatively small areas of 100 to 200 square miles as point source information on large-scale maps of one inch to one mile. These programmes proved particularly successful and demonstrated the need, following national or regional surveys, for immediate detailed studies as a means of delineating specific catchments, farms and the like for more conventional soil, vegetation and animal tissue surveys.

The patterns of elevated Mo concentrations reflect the occurrence of marine black shales ranging in age from Cambrian to recent deposits which may contain from one to 100 ppm Mo (Thomson *et al.*, 1972). In many of these areas, clinical hypocuprosis and/or low blood Cu values (perhaps indicating a sub-clinical problem) had been found. The wide distribution of low blood Cu values (hypocupraemia) in cattle in England is shown clearly in Fig. 3, in which over 1,700 herds were found to be "copper"-deficient, many of which were located in areas with molybdeniferous soils (Leech *et al.*, 1982).

Marine black shales similar to these are found in many parts of the world — and perhaps we have only started to explore their importance. As extensive systems of livestock farming are forced to become more intensive due to population pressure, natural limitations to food production of this type will gain increasing importance.

The disease swayback is a nervous disorder of lambs. It is characterised by lack of co-ordination of movement and paralysis of the back legs, and has been long recognised in many parts of the world. It is quite common in many parts of Britain, though only in certain years. The pathology shows myelin degeneration in the brain stem and spinal cord, and is associated with abnormal levels of copper in both the blood and tissues of mother ewes and their lambs. It can be prevented by giving copper supplementation halfway through pregnancy.

This disease has been studied by a large number of animal nutritionists, veterinary scientists, biochemists and pathologists — and it may be somewhat immodest for an environmental geochemist to offer a possible cause. However, an actual cause has been difficult to find and a copper-deficient diet would not seem to be the simple solution.

I visited New Zealand in 1970 and was very much attracted to the work at DSIR, where it had been shown that farm animals accidentally ingest appreciable amounts of soil (Healy, 1968). This was first noticed due to accelerated toothwear in sheep. I came back to Britain fired with enthusiasm and with Dr. Brian Alloway (now Professor Alloway, University of Reading), set up a programme funded by ARC to study the role of soil ingestion on livestock nutrition under British farming conditions.

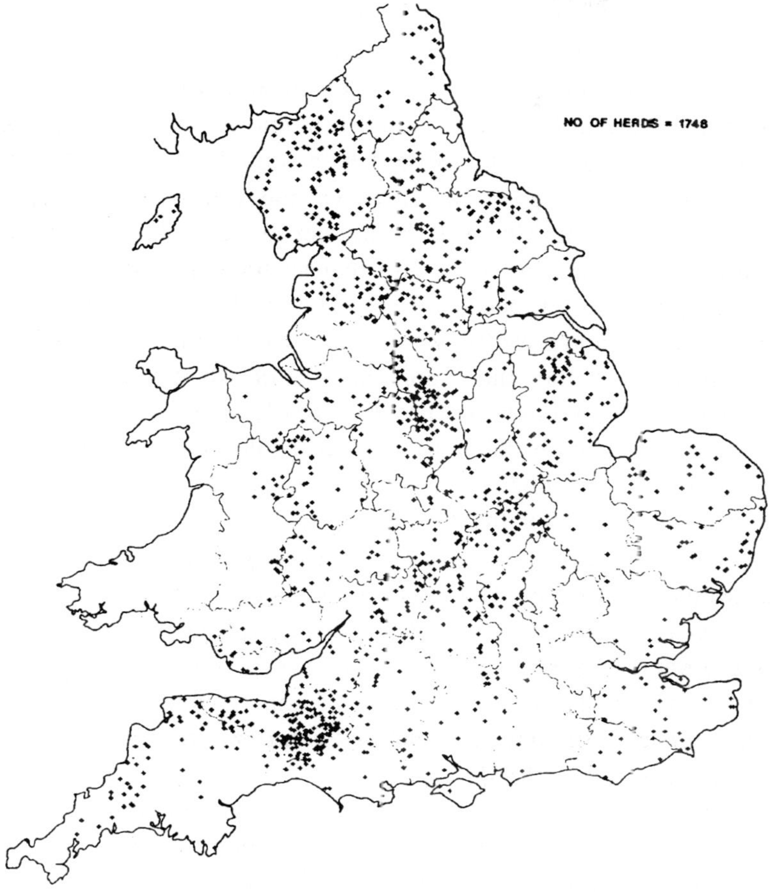

Fig. 3. Distribution of bovine hypocupraemia (blood Cu ≤ 0.5 mg/1) in England and the Welsh borders 1977–1980 (Leech *et al.*, 1982).

Titanium was used as a stable indicator of soil ingestion as soil contains several thousand parts per million Ti, while normal pasture plants contain less than 10 parts per million. By measuring concentrations of Ti in the animals' faeces, the amount of soil ingested can be calculated. We found that between two and 20% of the dry matter intake of grazing cattle was soil, and up to 30% or 40% in sheep as they graze closer to the ground. This led to an examination of the actual ingredients in soil that may influence animal metabolism and nutrition.

The trace nutrient Cu was a natural first choice and, in collaboration with Dr. Neville Suttle of the Moredon Institute, soil was fed to sheep whose liver reserves had been run down and were on a Cu repletion diet. It was found that 10% soil in the diet reduced Cu absorption and utilisation by as much as 50%, and that as little as two per cent soil ingestion significantly reduced Cu use. The mechanism for this was not immediately clear and it was not known whether the soil particles occluded Cu in the diet, whether they supplied Mo, or whether the effect was due to some as yet unknown process (Suttle *et al.*, 1975).

In some very simple laboratory experiments, soil was shaken with rumen liquor in a test tube. The large amounts of Fe coming into solution from this soil gave a valuable clue.

This work was repeated on a larger scale using artificial rumen systems at the Moredon Institute. The released Fe is a powerful Cu antagonist, and it was proposed that accidentally ingested soil provides a source of large amounts of Fe in the rumen of sheep. This Fe interacts with dietary Cu and reduces the utilisation of Cu. It would thus seem that this mechanism could make a significant contribution to the aetiology of swayback (Thornton, 1983).

I would suggest that we have found one of the main causes of this economically important disease. Swayback is found in years with warm winters and very little snow cover. Indeed, MAFF put out a warning to farmers in years when the number of days with snow cover are low. In warm weather, sheep are left to fend for themselves on poor pasture with little or no extra feed. This is when soil ingestion is at its highest and when supplementary Cu is needed by the pregnant ewe. At a meeting of the British Society of Animal Production and the British Veterinary Association in 1982, I was able to present this case, and I am not aware of it being disputed (Thornton, 1983).

The next question addressed was whether ingested soil contributed significantly to the supply of nutritionally essential elements for the grazing animal under UK conditions and, in particular, the trace element cobalt. Cobalt is an essential constituent and, indeed, comprises some four per cent of vitamin B_{12}, the anti-pernicious anaemia factor of liver which is produced by the synthesising abilities of the ruminants symbiotic gastro-intestinal flora (Underwood, 1962). Nutritional deficiencies of Co result in severe cases of extreme emaciation and listlessness. These may lead to death in both cattle and sheep; milder forms are usually described as an ill-defined and often transient state of ill thrift and respond rapidly to cobalt supplementation (Underwood, 1962). Cobalt deficiency in sheep has been recorded in several areas in the UK underlain by arenaceous course-grained soil parent materials and by acid igneous rocks which contain only very small amounts of cobalt. It is normally treated by the direct oral administration of "cobalt bullets" to the animal or by the treatment of deficient pastures with cobalt sulphate.

There is 100 to 1,000 times more Co in soil than in pasture herbage, and potentially, a little soil could provide the dietary requirements of the animal, irrespective of how much is present in plants.

Further collaborative studies between Imperial College and the Moredon Institute, with particular research input by Brebner, showed that not only was Co extracted from soil in the sheep's rumen, but it was also synthesised into Vitamin B_{12} as required by the sheep (Brebner, 1986). The diagnostic practice of MAFF for the past 20 or more years has been based on the analysis of pasture herbage for Co. On the evidence of this research, it is suggested that this may not have been appropriate as sheep directly ingest appreciable quantities of Co in soil (Brebner *et al.*, 1986).

I must emphasise that we still have a lot to learn about the practical significance of soil ingestion to livestock production and animal health. Unfortunately, research aimed at increasing agricultural production is nowadays not so fashionable in the UK. This is an indirect result of the Common Agricultural Policy and the move towards less intense agricultural systems.

Geochemistry and Pollution

Now, I would like to consider man's activities which have resulted in a change in the chemistry of the surface environment, and that interact with, modify and even mask natural geochemical processes and pathways. I will focus on the metals, lead and cadmium.

Under my direction, the Applied Geochemistry Research Group has made a series of detailed studies into the sources, distribution, chemical forms, pathways, exposure and impacts of lead both to farming and to public health. Funding for this continuing programme of work has been provided by the research councils, government and industry.

It has been recognised for several centuries that lead is a poison and there is a long history of human exposure to lead in food and drink — for example, in ancient Rome — in amounts which must have caused frank lead poisoning (Gilfillan, 1965). Over the last century, many cases of lead poisoning at work have been documented. However, acute lead poisoning in the community is rare today and is usually associated with exposure to specific localised sources. In recent years, however, there has been growing concern that lead in the body resulting from general environmental exposure may be harmful at concentrations which fall short of those at which clinical symptoms may appear. Particular attention has been paid to the possible effects on the intelligence and behaviour of young children (Landsdown *et al.*, 1974; Needleman *et al.*, 1979).

Of course, lead is present in large quantities in the earth's crust and, in the course of weathering, some reaches the soil, air and living organisms through natural processes. But this natural distribution is insignificant when compared with the distribution which results from man's activities. Man has used lead for thousands of years. Its ore is plentiful, accessible and simple to smelt. The metal is heavy, very pliable and very resistant to corrosion and weathering. It is easily melted down and re-used. These properties account for its widespread traditional uses, for instance in building, plumbing (to which it gives its name), printing, shooting and fishing — many of which continue today.

The use of lead shot is an example. In the UK, 120 million cartridges are sold each year, of which 45% are used in clay pigeon shooting, and 55% in game and wildfowl shooting and pest control. Of the 3,000 tonnes of lead dispersed into the environment in this fashion, some is ingested by cattle and some by birds. This can lead to lead poisoning. Studies on a clay pigeon shoot on farmland in Oxfordshire that had been used for some 30 years showed that all shot fell within 200 m of the shooting traps/stands. This resulted in accumulations of up to 1,000 pellets/m^2 in the top five centimetres of soil. In one legal case, this certainly resulted in deposition of lead shot onto a crop of maize used for silage; a test tube full of shot was recovered from a cow's rumen that had been fed this silage.

Even where substitutes are now preferred or required, artefacts from earlier times survive and are still in widespread use. Since the Industrial Revolution, other properties and uses of lead have been discovered and widely exploited. In its metallic form, lead is uniquely suitable for use in heavy electrical industries, insulation, radiation shielding and battery manufacture. Various compounds of lead have been, and at the time of this lecture were still, used in paints, plastics, ceramics, crystal and petrol.

The cumulative effect of centuries of use is that lead is now one of the most widely dispersed of the environmental pollutants. Moreover, so far as is known, there is no innocuous form into which it can be converted in the environment. Once in the environment, lead and its compounds do not move readily through natural pathways such as the ocean bed, and they have a long environmental residence time compared with other pollutants, particularly in soil. Thus, not only is there widespread human exposure to lead today, but future generations too will be exposed to lead which is already in the environment, and which is being added to all the time.

The geochemical map for lead in England and Wales (Fig. 4) shows patterns of elevated lead concentrations that largely reflect our long history of mining and smelting (Thornton, 1980, 1996b). They also reflect some recent or present-day activities, including an anomalous pattern around the world's largest zinc/lead smelter at Avonmouth, as well as contaminated

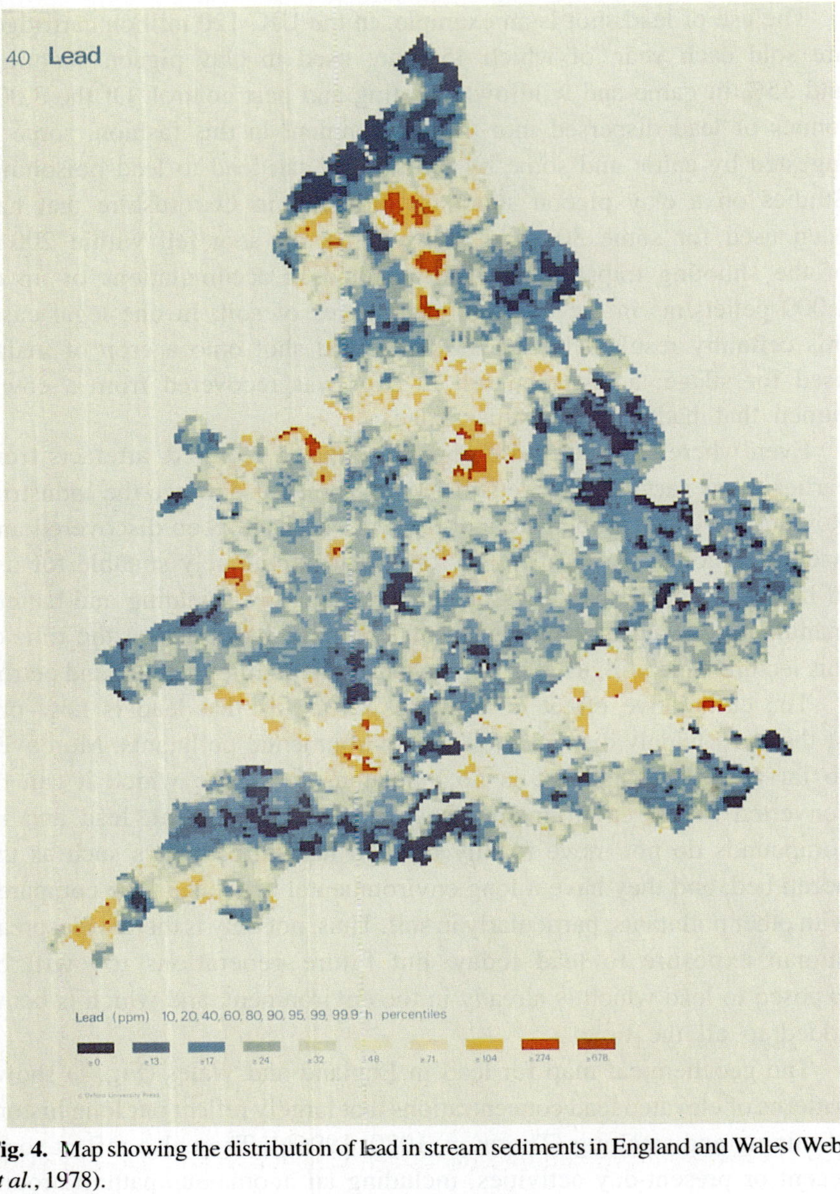

Fig. 4. Map showing the distribution of lead in stream sediments in England and Wales (Webb *et al.*, 1978).

stream beds around Birmingham, the UK's second largest concentration and its associated heavy industry. In all, some 4,000 km^2 of contaminated land are delineated.

The question must be asked as to what impact this historical legacy and present-day pollution has in terms of agriculture and human health. We have looked at mine wastes, soils, plants, garden produce, waters, dusts, air, bovine and human blood and hair, and we are currently concerned with the forms and chemistry of lead that may influence the metal's solubility, bioavailability and pathways into the food chain. In Derbyshire, there is an area of some 250 km^2 of contaminated soil with several hundred to several thousand ppm Pb and as much as even one, two and three per cent Pb near old mines and smelters (Colbourn & Thornton, 1978). Occasionally, clinical Pb poisoning in cattle and sheep is recorded when animals stray onto old spoil tips, though this is far more common when cattle chew old batteries and paint in the farmyard.

Over a two-year period, a detailed study was made of lead on 11 farms in collaboration with the Veterinary Investigation Service of MAFF. As indicated in Fig. 5, only a little lead was taken up into the pasture during the grazing season. However, a blood lead survey undertaken with help from a team from St Mary's Hospital led by Professor Donald Barltrop showed blood lead to reflect the lead in soil (Fig. 6). Again, this was shown to be a clear case of the impact of soil ingestion. There were, however, no obvious health effects to the cattle, no apparent effects on growth or milk production, and no appreciable influence on the composition of the milk.

Human Exposure to Lead

In the late 1970s, a collaborative study involving the Paediatric Unit, St Mary's Hospital Medical School, and AGRG showed that when households in old mining villages in Derbyshire were grouped according to the Pb content of their garden soils, Pb in the blood and hair of two- to four-year-old children increased with that in soil and housedust (Table 1) (Barltrop *et al.*, 1975).

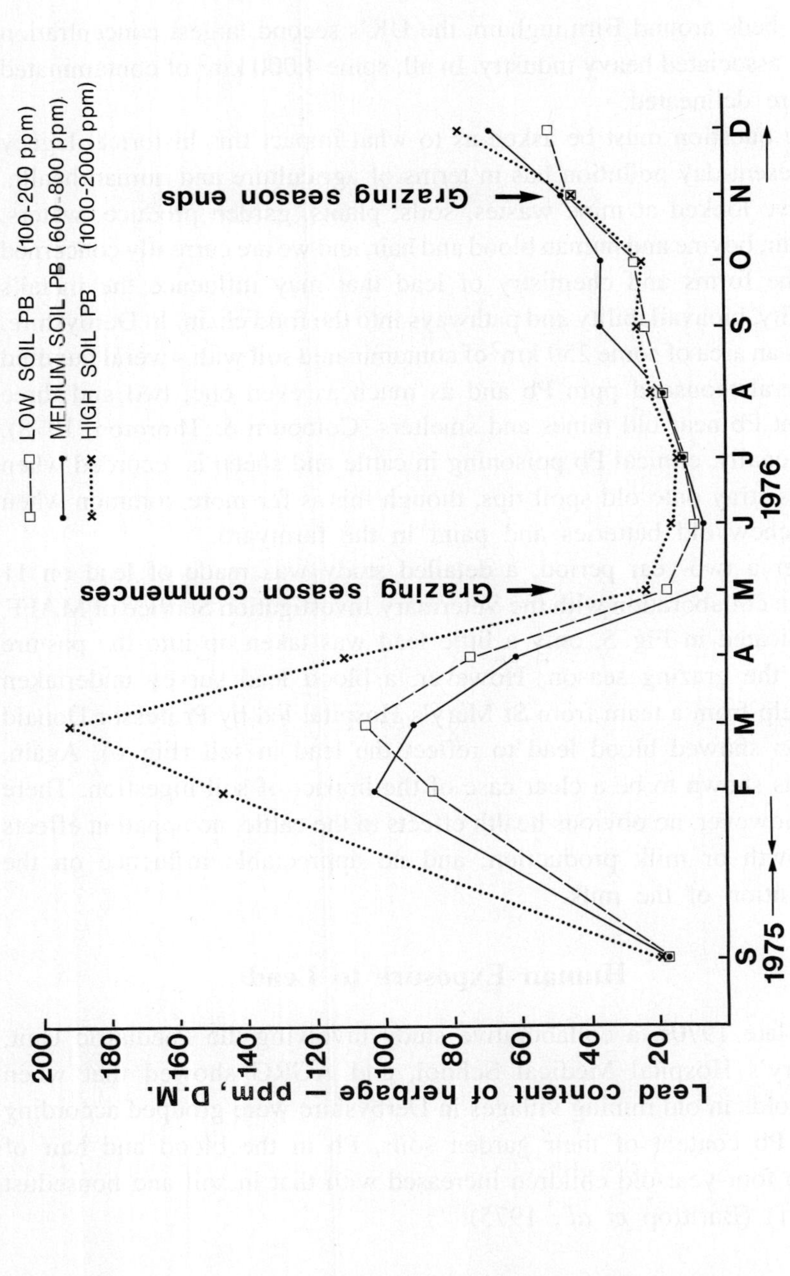

Fig. 5. Uptake of lead by pasture herbage from soils of different lead contents.

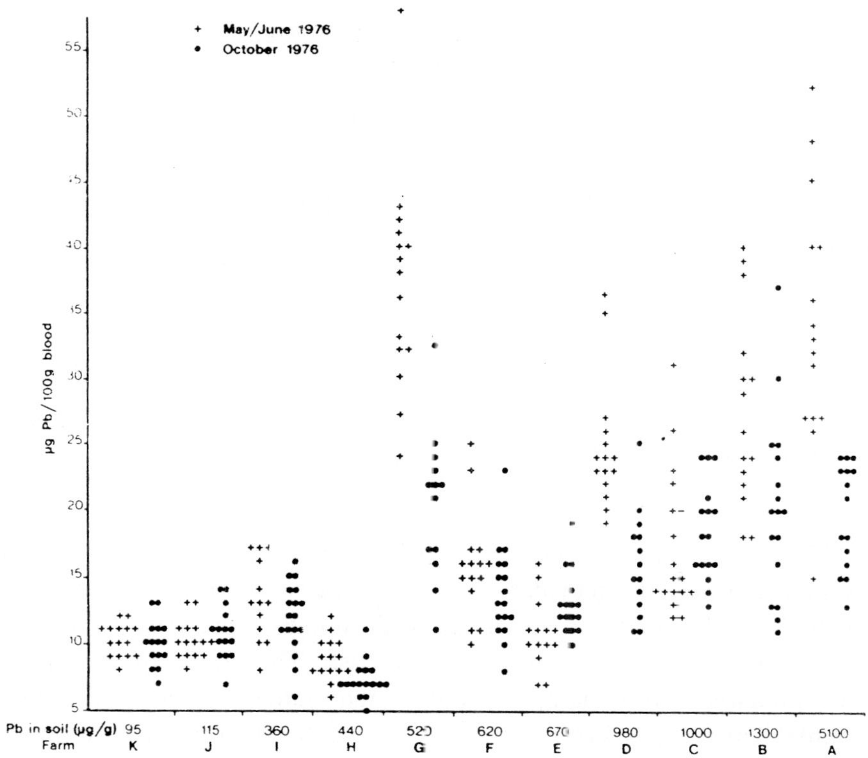

Fig. 6. Blood lead concentrations in herds of cattle grazing land with different lead contents.

However, none of the Pb values in children were high enough to be considered hazardous at that time, even though the amounts in soil and dust were extremely high, peaking at 2.8% and 2.5% Pb, respectively. The children's Pb status did, however, reflect that of the environment, and in the absence of evidence linking the human burden with Pb in foodstuffs or water supplies, it was suggested that the major pathway of Pb into the children was through inhalation and involuntary ingestion of dust particles by hand-to-mouth activity.

Table 1. Mean blood and hair lead content in Derbyshire villages.

Soil lead	Blood Lead (μg/100ml)				
	Child	Mother	Hair (p.p.m.)	Soil (p.p.m.)	House Dust (p.p.m.)
Less than 1000 p.p.m. (N = 29)	20.7[a,b]	14.1[c]	7.7[f]	420	531
1000–10,000 p.p.m. (N = 43)	23.8[a]	18.7[c,d]	10.5[e]	3390	1564
Over 10,000 p.p.m. (N = 10)	29.0[b]	14.8[d]	20.2[e,f]	13,969	2582

a, *a*; *b*, *b*; etc. significantly different, two-tailed *t*-test, $P < 0.05$.

More recently, the village of Winster and its surroundings were re-examined in some detail to identify possible sources and routes of exposure to Pb. Particular emphasis was placed on the homes of pre-school children, the members of the community most at risk. The results of this study are summarised in Table 2 and confirm that garden soil and housedust samples have highly elevated lead levels (Cotter-Howells & Thornton, 1991; Cotter-Howells, 1991). Handwipe samples from children also have relatively high lead concentrations, suggesting that elevated levels of lead are transferred to the child by the soil-dust-hand-mouth pathway. However, this is not reflected in their blood lead concentrations, which are within normal UK ranges and less than that predicted by some lead exposure models. SEM X-ray maps show galena weathering directly to cerussite ($PbCO_3$) in the soil environment, which is consistent with the pH (6.6–8.2) of these soils (Garrels & Christ, 1965). Many lead-rich grains were composed of pyromorphite ($Pb_5(PO_4)_3Cl$), which is a stable soil-lead mineral (Nriagu, 1974). The mode of formation of this mineral is not clear. It may have formed in the mineral vein and then been incorporated into the soil or formed within the soil as a result of further weathering of galena.

Table 2. Lead in environmental samples, handwipes and children's blood in Winster, Derbyshire.

	Winster		
	n	GM	Range
Garden soil (Pb μg g^{-1})	42	7,140	2,400–22,800
Vegetable soil (Pb μg g^{-1})	11	9,580	2,235–22,160
House dust (Pb μg g^{-1})	45	1,560	462–6,240
Lead loading (Pb μg m^{-2})	44	275	14.2–3,560
Handwipes (μg Pb per pair of hands)	10	13.1	3.8–39.5
Children's blood (Pb μg dL^{-1})			
1–3 year olds	3	6.9	6.2–8.3
1–8 year olds	10	9.4	6.2–20.7

Pyromorphite is only very slightly soluble ($K_{sp} = 10^{-84.4}$) (Nriagu, 1973), less so than galena ($K_{sp} = 10^{-27.5}$) (Smith & Martell, 1976). It could contribute to a low bioavailability of lead in these soils, below that of urban soils where more soluble forms of lead are present (for example, lead chloride and bromide). A low soil-lead bioavailability would reduce the uptake of lead into the child so that despite increased lead intake via the soil-dust-hand-mouth route, blood lead concentrations could remain within normal UK ranges.

None of the children's blood lead values exceeded 20 ug/dl, which is below the 25 ug lead accepted as the threshold for medical follow-up today. However, this threshold may well come down to 15 ug/dl and then the situation may have to be re-assessed.

Geochemistry of the Urban Environment

Of course, the majority of the population do not live in old mining villages but in cities and towns — the urban environment. I have coined the term "urban geochemistry" to describe our research activities concerning the role of the geochemists at the interface of environmental geochemistry and urban pollution.

In 1981, I was invited by the DOE to undertake a national study of lead and other metals in urban dusts and soils. The aim was to determine the normal ranges of metal concentrations to which the UK population might be exposed to in and around their homes, with a strong emphasis on the role of dust as a source of lead to the young child.

The importance of dust ingestion as a source of lead exposure had already been highlighted by the Royal Commission on Environmental Pollution (1983) under the chairmanship of Professor Sir Richard Southwood.

UK National Survey of Lead in Dusts and Soils

A nation-wide survey of metals in dusts and soils commissioned by the Department of the Environment was undertaken in 1981/82 in 53 locations in England, Scotland and Wales (Culbard *et al.*, 1988; Thornton *et al.*, 1985). Cities, towns and villages were selected to reflect differences in geographical location/geology), degree of industrial/urban development and population distribution. Seven London Boroughs were included. Known "geochemical hotspots" in areas of past mining and smelting activity were added to reflect likely acute exposures.

This survey showed that:

1. Concentrations of Pb in dusts and soils were sufficiently large to give rise to concern that a proportion of urban children were likely to be exposed to significant amounts of Pb.
2. Overall, 10% of homes had in excess of 2,000 μg/g Pb in housedust; 5% of garden soils were above this level.
3. London had significantly higher levels of Pb, with 18% of homes and 10% of soils exceeding 2,000 μg/g Ob.
4. "Geochemical hotspots" showed significant Pb contamination within the home. In Winster, Derbyshire, for example, 44% of housedusts had more than 2,000 $\mu g\ g^{-1}$ lead and 93% of soils exceeded this level, with 1% Pb or more in several gardens.

Table 3. Geometric mean lead concentrations (μg/g^{-1}) of dusts and soil in and around houses of various age.

	Lead Concentration		
	Dust		Soil (0–5 cm)
House Age	Vacuum	Road	
Pre-1870	982	1,483	1,146
1870–1919	1,874	1,607	1,014
1920–1939	619	681	368
1940–1959	433	472	292
1960–1986	241	316	131

A questionnaire for each household was used to provide information on the factors that might lead to the accumulation of lead in housedust, including age of property, distance from roads, road type, occurrence of old lead paint and recent history of decorating activity. In Brighton, England (Davies & Thornton, 1987), lead in both dust and garden soil increased with increasing age of the property (Table 3).

This would suggest that internal sources of lead are more important in normal urban households than external sources, except in old mining areas where the huge amounts present in contaminated soils are the controlling factor influencing that found within the home, and here we must remember that housedust can contain as much as 50% soil.

The national study clearly confirmed that lead contamination is widespread in both dusts and soils in Britain, and it established an urgent need to assess the relative importance of lead-rich dust and soil as a source of lead exposure to the young child, with a view to providing reliable quantitative information. Subsequently, in 1982, a comprehensive study of 97 two-year-old urban children was carried out in inner-city Birmingham. It was designed to provide quantitative information simultaneously for Pb intakes via all identified pathways (Davies *et al.*, 1990). This multi-disciplinary programme involved collaboration between scientists at Imperial College (environmental assessment), University of Southampton (blood lead determinations), University of Birmingham

Table 4. Intake and uptake of lead by inhalation and ingestion by 2-year old children.

Source	Units	Birmingham Study	Royal Commission (1983)
Inhalation			
Air-lead concentration outdoors	μg m^{-3}	0.43	1.00
Air-lead concentration indoors	μg m^{-3}	0.26	0.75
Time spent indoors	h day	23.25	20
Volume of air respired	m^3	6	6
Lead intake from air	μg day^{-1}	1.56	4.8
Percentage absorption in lungs		70	70
Lead uptake from lungs	μg day^{-1}	1.1	3.4
Ingestion			
Dietary lead intake	μg g^{-1}	23	52
Dust-lead concentration	μg g^{-1}	424	1400
Lead intake from dust	μg day^{-1}	42	140
Total intake by ingestion	μg day^{-1}	65	192
Percentage absorption in gut		53	53
Lead uptake from gut	μg day^{-1}	35	102

(children's behaviour), MAFF Food Science Division (duplicate diet study) and Department of the Environment (co-ordination and statistical evidence). This provided detailed measurements which allow for the calculation of total Pb uptake based on Pb intake from air, dust and diet (Davies, 1987) (Table 4). With an average air Pb concentration of 0.26 μg/m^3, an average time spent indoors of just over 23 hours a day and an assumed volume of respired air of 6m^3/day, the Pb intake from air would be 1.5 μg/day. Assuming a combined retention and absorption factor of 0.7, Pb uptake would be just over 1 μg/day in addition to the 23 μg Pb/day from the diet (food and beverages). Assuming that lead from dust and from the diet are equally bioavailable, and that 53% of the lead ingested reaches the bloodstream from the gut, the uptake by ingestion would be 35 μg Pb per day. The total estimated uptake of 36 μg/day is only about one-third of that calculated by the RCEP (1983) and is much closer to that estimated for two-year-old children in the US (EPA, 1986). However, given the uncertainties surrounding the figures for dust ingestion and absorption by the gut, both the estimated proportions and total amounts of Pb from dust and the diet must be treated with caution.

Each child's behaviour was recorded on video in its own home over a single four-hour session. Samples of the video were analysed using a detailed set of behaviour categories for mouthing activity and hand contact. A 30-item questionnaire about mouthing behaviour was given to the primary carer (usually the mother). Using the behavioural data and the environmental, dietary and blood lead measurements, a regression model was developed relating blood lead to the other measured parameters (R2 = 35%):

$$\text{Log PbB}(\mu g/100mL) = 0.55 + 0.10 \log x\ 1 \\ + 0.141 \log \text{PbW}\ (ugL-1) + 0.07S,$$

where x1 is the dust lead loading multiplied by the rate for hands touching all objects (including the floor), PbW is the water lead concentration and S is the dichotomous variable equal to one if either parent smoked and zero if both were non smokers. Non-significant regression coefficients and only marginal improvements in R2 values resulted from the addition of air or soil-lead concentration, total handwipe lead, dietary lead or any combination of these.

The next question we had to answer was where does this lead within the home come from. And, of course, elevated levels of indoor dusts result from contributions made from a variety of possible sources. One analytical technique which showed considerable promise for addressing the question of lead source attribution combines scanning electron microscopy (SEM) and energy dispersive X-ray analysis (EDX) (Johnson *et al.*, 1985). SEM, in tandem with EDX, is capable of providing both compositional and morphological data on individual lead-bearing particles present in environmental dusts. While the techniques proceed on a particle-by-particle basis, the time efficient analysis of significant numbers of particles becomes possible by running the SEM and EDX analysis under computer control. Variations in size, shape and composition make it possible to differentiate between lead-bearing particles from different emission sources. For example, Pb particles in fresh auto exhaust are generally characterised by the presence of the elements Br and Cl (Watt & Johnson, 1987).

To develop a source library suitable for the apportionment of urban particulate lead, material was sampled from a variety of potential sources of lead (Hunt *et al.*, 1988). These included fresh vehicular exhaust, lead-based paint, primary and secondary lead smelters, a waste incinerator, electrical solder, lead metal and oxide reference particles, road dusts and urban soils. The latter two sources were selected as being descriptive of the transport pathway into the home (Hunt *et al.*, 1991). In all, over 8,000 generic source particles were analysed. Typically, a particle class balance receptor model approach is used. In the "model" we have proposed (Hunt *et al.*, 1991), we suggest that the resolution provided by the 101 class scheme is sufficient to avoid the use of statistical apportionment procedures.

The Nationwide Reconnaissance Survey found that house dust levels in London were among the highest in the UK. To apportion the sources of lead in homes in Richmond, 16 house dusts were selected for analysis (Hunt *et al.*, 1992). These were chosen from each of five house age categories: pre-1870, 1870–1919, 1920–1939, 1940–1959 and post-1960. Each house dust had been found previously to have a lead concentration close to, or below, the mean value for house dusts in London (1,010 mg kg^{-1}). Some 17,000 fields of view were examined, and over 66,000 high backscattered electron thresholded particles were analysed, of which 10,175 were lead-bearing. The analysis results have been presented in detail elsewhere (Hunt *et al.*, 1993).

In summary, while paint was found to be the predominant source of lead in larger particles, important inputs from dust and soil were found in the finer fractions.

I am frequently asked where I stand in the lead debate. I have been impressed with the results of research into neurotoxicology, where it has been shown that lead can affect neuronal development in the foetus at levels of exposure found in the environment — that lead can cross the blood-brain barrier; that lead may alter the organisation and function of cortical wiring in the brain — all of which would help to provide some information on the mechanisms by which lead exposure, even at low or moderate levels, during foetal development and early postnatal period may result in cognitive impairment. Longitudinal studies into the effects of lead on young children in the US and in Australia mostly confirm adverse effects of lead exposure on indices of mental development and other measurements regarding learning and behaviour. Results from studies in Edinburgh have, for the first time, confirmed similar adverse effects on children's learning capacity in the UK (Fulton *et al.*, 1987). My present opinion is that we should continue to view the lead debate with some caution; those in government would probably use the word prudence. The evidence is moving in one direction only. We have amounts of lead in our home environment today that are potentially toxic to the young child — that can possibly impair neurological development and cognitive function. This lead is around for possibly centuries to come. We must learn to reduce exposure and must ensure that further inputs to our soils, home and foodstuffs are kept to a minimum.

Cadmium

Now, I will turn briefly to the metal cadmium. Cadmium is another non-essential element present as a contaminant in food and water. Concern about its effect on the general population exposed to low concentrations over long periods of time has been raised, in part, because of its steadily increasing consumption and release into the environment in the last 20 years, and partly because of the outbreak of "Itai-Itai" disease in Japan, for which Cd has been shown to be one causative factor.

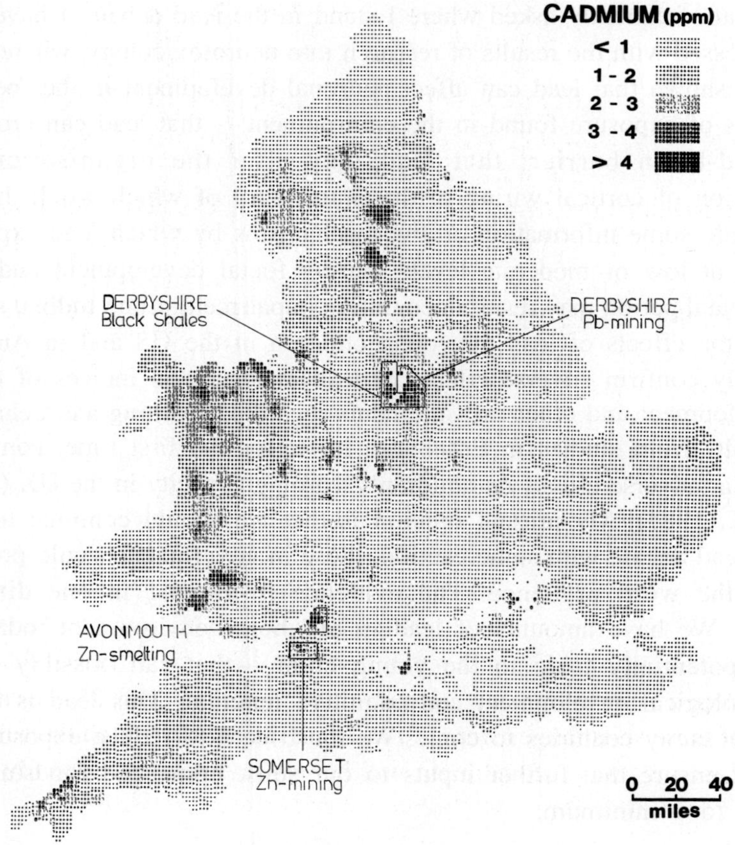

Fig. 7. Map showing the distribution of cadmium in stream sediments in England and Wales (Webb *et al.*, 1978).

The geochemical map for cadmium again focuses attention on areas contaminated by mining and smelting (Fig. 7). A comprehensive environmental health study was planned around the village of Shipham, Somerset, where zinc was mined as the ore smithsonite ($ZnCO_3$) over the period 1700–1850. The ore, with appreciable amounts of Cd as a guest element, was washed in ponds and calcined within the village. Collaborative investigations involving the national and local governments, the Westminster Medical School and AGRG showed that:

1. Sixty per cent of household garden soils had over 60 ppm Cd, greatly exceeding the levels in polluted paddy soils associated with the well-documented "Itai-Itai" disease in Japan.
2. Household dust samples averaged 26 ppm Cd and 2300 ppm Zn.
3. From studies on metals in locally grown vegetables and diets, an average human intake of 200 ug Cd per week was calculated, compared with the average UK intake of 140 ug Cd per week; individual intake rarely exceeded World Health Organisation provisional tolerable weekly intake of 450–500 ug Cd_3.
4. Through health inventories and biochemical tests on 548 residents of Shipham and 543 controls from a nearby uncontaminated village, only slight differences attributable to Cd were found (Morgan, 1988; Strehlow & Barltrop, 1988; Thornton, 1988).

In this unique situation, it is possible that large amounts of Zn and Ca present in the environment are affording some degree of protection against possible adverse health effects of cadmium. The soils in the village are developed from calcareous dolomitic conglomerate parent material. The pH of these soils ranges from 7.0 to 7.8, a range over which the solubility and bioavailability of cadmium will be low and uptake by locally grown vegetables small (Moir & Thornton, 1990). Thus, the local population are afforded some degree of natural protection. However, many of the older residents who had lived in the village for a long time were not willing to participate. Hence the very people who might have been at risk were largely left out of the medical study.

I would forecast that at some time in the not too distant future, Shipham will be re-evaluated and with the present more rigorous environmental standards, the outcome could be quite interesting.

Further Studies in the Urban Environment

More recent pioneering research in urban geochemistry has concerned

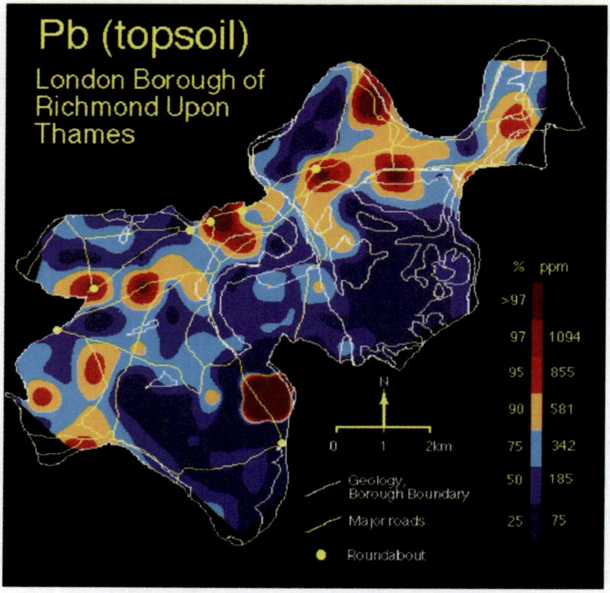

Fig. 8. Distribution of lead in surface soils (0–15 cm) in the London Borough of Richmond-on-Thames.

the identification of metal sources in towns and cities, and the degree to which they change or mask natural geochemical signatures in the surface environment (Kelly *et al.*, 1996). Richmond-upon-Thames, one of the 32 boroughs within the Greater London area, was chosen as a residential suburb with little history of industrial activity. Systematic sampling and analysis of surface soils over an area of 56 km^2 has clearly shown a difference in the distribution of lead between those parts affected by urban development and the remaining areas of parkland and open space (Fig. 8). The parts which show the highest concentrations of lead generally occur close to road junctions and roundabouts on main roads with high traffic density. Although unleaded petrol now accounts for 80% of sales, a legacy of contamination from past years will remain far into the future, as lead has a very long residence time in soil. Parallel studies have shown that the use of platinum compounds in catalytic converters has resulted in detectable concentrations of this metal on heavily trafficked roads (Farago *et al.*, 1998).

A comparative study of the city of Wolverhampton, with a long history of industrial activity, has shown contamination of zinc, iron and, to a lesser degree, cadmium and lead in the vicinity of old industrial sites (Kelly *et al.*, 1996). Agricultural land on the outskirts of the city has significantly lower concentrations of all heavy metals studied. This study has clearly demonstrated that heavy metal concentrations in urban soils are strongly influenced by past industrial use and must be taken into account in planning future development.

Geochemistry and Wildlife Nutrition

Further to the successful application of geochemistry to agriculture and the recognition that geochemical maps could delineate areas in which food crop and farm livestock production and health could be influenced by trace element deficiencies and excesses, I was approached by the Peoples Trust for Endangered Species to see if our research could be applied to wildlife conservation areas in Kenya. Here, a potential problem had been recognised by the Department of Wildlife within the Lake Nakuru National Park. It was designated as the first sanctuary for the black rhinoceros, an animal in danger of extinction as a result of poaching. Here there is a history of mineral deficiencies of ranched cattle, and both the impala and water buck show signs of possible deficiencies. The impala show weakness and stiffness of the back legs and the water buck, emaciation and loss of coat colour.

The problem is that wildlife are now enclosed in what can be described as island parks. Instead of being able to roam over very large areas, the animals are now dependent on herbage and browse plants in relatively small areas. Lake Nakuru National Park is only 160 km^2. Some 20 black rhinoceros have now been moved into the Park, which is also the home of the relatively rare Rothchilds giraffe.

A reconnaissance geochemical survey was carried out in the Lake Nakuru Park to establish the mineral status of soils and selected plant species and the results related to the health of the animals (Maskall & Thornton, 1989). Soil and plant samples collected on a one-kilometre grid were analysed for 25 elements using Inductively Coupled Atomic Emission Spectrometry and blood samples analysed for copper and vitamin B_{12}.

The total concentrations of copper and cobalt in soils were low (Figs. 9a & 9b), a geochemical feature shared by many of the Rift Valley soils derived from volcanic ash sediments and other volcanic rocks. Total soil selenium and phosphorous levels were also relatively low. Grass species contained higher levels of copper and cobalt and lower levels of selenium compared to browse plants. Molybdenum levels in all plants reached relatively high values and the availability of this element appeared to increase in wetter soils of high pH near the lake shore (Maskall & Thornton 1991, 1992). Over 30% of impala sampled had a blood copper level below that regarded adequate for domestic animals. The relatively high molybdenum content of grasses and browse plants is believed to contribute to possible copper deficiencies of impala and water buck in the park. The lack of data on the mineral requirements of other wildlife species prevents an assessment of the risk of deficiencies to these species at the present time.

The question is then raised as to whether we can do something about this — and the answer is yes. Wildlife already eat soil — the term "salt lick" has been applied to those soil exposures where they intentionally visit. Game wardens and farmers have found that most species, including rhino, will eat salt and other minerals spread on the surface of the soil. They will also take supplementary minerals put out for ranched cattle. As an interim measure, I have made recommendations to the Kenya Department of Wildlife Conservation and Management that mineral salts with Co, Cu and Se are made available in Nakuru Park. This approach to wildlife nutrition is new and this is the first time that geochemistry has been applied. There are, of course, numerous other situations when this type of approach could be applied. Last year, I visited the Oman

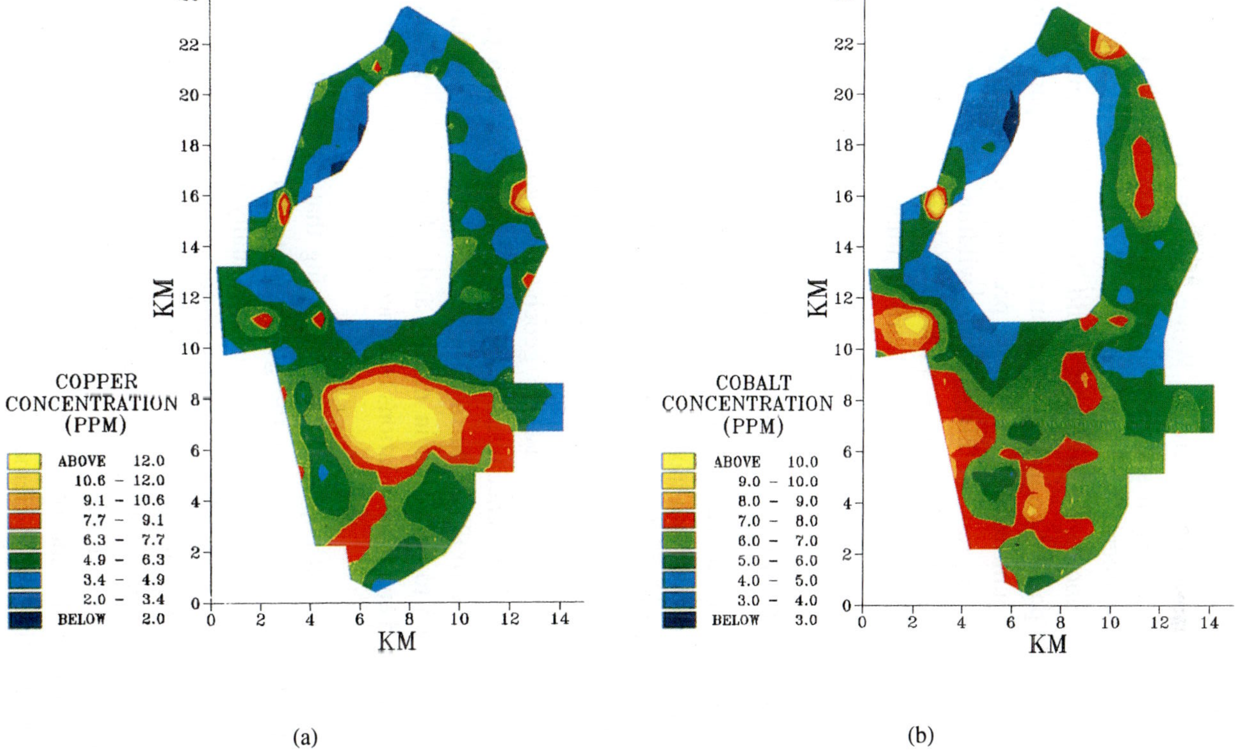

Fig. 9. (a) Lake Nakuru National Park: copper concentrations in surface soils (0–15 cm); (b) Lake Nakuru National Park: cobalt concentrations in surface soils (0–15 cm).

to advise the Diwan of the Sultan on a similar approach, in part, of the Jiddat Al Harasis Desert where the White Arabian Oryx, previously extinct in the wild, has been successfully re-introduced mainly from a zoological park in southern California. Animals thrive better in one part of the desert than another. Again, this could be a mineral problem.

With the world population increasing at an alarming rate, particularly in the developing countries such as Kenya, pressure on land, especially farmland, continues to grow. In this situation, wildlife will be restricted to relatively small protected areas — such as Nakuru — and, as on a farm or on a ranch, it is necessary to learn to understand and satisfy their dietary needs.

Geochemistry and Endemic Disease in Man

Attempts to relate trace element deficiencies and toxicity in man to the geochemical environment have only been partly successful, and have presented the scientist with some difficulties (Thornton, 1987). There is an obvious temptation with the availability of geochemical maps on the one hand, and maps showing the distribution of human disease on the other, to attempt to fit diseases to geochemical patterns as causes of disease. The problems encountered have been discussed in a paper entitled *Epidemiology for Geochemists* (Shaper, 1979). Few relations have, in fact, been shown to be causal in nature, and it has only proved too easy to put forward empirical relations that meet with very little scientific assessment.

In modern society, it is perhaps not surprising that the health of the population does not often reflect the geochemistry of local soils and rocks. First, sources of human food are continually widening so that the overall diet usually contains foodstuffs grown or produced on a range of soil types with differing geochemical characteristics. Secondly, modern diets, particularly in the Western world, contain a variety of foods. Hence the mineral imbalance in one type of food may be offset by another. Thirdly, man is at the end of the food chain so that gross deficiencies or toxicity affecting trace element levels in plant and animal tissues would generally have been modified in order to maintain crop and livestock yields. Finally, technological development in food production and food processing, such as the refining of sugar, milling of wheat, canning and use of food additives, can result in gains and losses in trace elements in food, which can erode the direct relationship between man and his natural geochemical environment.

However, there are some obvious exceptions which include:

1. The well-documented relation between F content of waters and the incidence of human fluorosis (and inverse relation with dental caries).
2. Relation between the prevalence of endemic goitre and I levels in food.
3. The prevalence of the Se-responsive Keshan Disease and Kaschin-Beck Disease in areas of low Se status in parts of the People's Republic of China.

For several years, I have been working on a collaborative research programme with Professor Yetang Hong of the Institute of Geochemistry in Guyang, supported by the Royal Society and the Chinese Academy of Sciences, to examine some of the relationships between geochem-istry and human health in China. Part of this programme is concerned with fluorosis which affects some 30 million people. It is recognised that endemic fluorosis may arise as a result of F-enriched surface waters, F-rich groundwaters, F-rich hot spring waters and the burning of F-rich coal (Zheng & Hong, 1988). The latter situation is one of appreciable interest and affects some one million inhabitants in Zhijin County in Guizhou province. Originally ascribed to the ingestion of F-rich water supplies (Lyth, 1946), it is now recognised that the problem arises as a result of burning coal which may contain as much as 2000 ppm F in stoves with no ventilation to the exterior environment, above which locally grown food crops such as corn and vegetables are dried and contaminated by smoke. The main crop, corn, has an average F content of 0.4 ppm (dry matter) prior to drying, but contains 10 to 100 ppm F after drying and is the cause of heavy dental mottling in almost all the population and crippling skeletal fluorosis in 10%. This, indeed, is a clear case of a geochemically related disease. Ongoing collaborative research between the Institute of Geochemistry, the Chinese Academy of Sciences and Imperial College continues to investigate the influences of geochemistry on fluorosis and Se-responsive diseases.

Conclusions

I have attempted to explore and describe some of the more applied aspects of my work and that of my students and colleagues over the past 25 years. I would like to conclude by addressing some general principles. Geochemistry has, and will, continue to influence plant, animal and human health, to influence ecological communities, and to influence human communities. But man, with his increasing numbers, will continue to change the use, and the chemistry, of this extremely complex and diverse environment that we share. Many of the natural interactions between components of the geosphere and the biosphere will be upset as many of the geochemical cycles are modified.

Some of these changes will be for the good. We can modify the soil to produce more food. We can modify animal and human diets to reduce endemic disease and to improve health and well-being. But we are increasingly contaminating and polluting the various components of the environment and we are far from understanding the effects of this.

To what degree are the different components of the system geosphere/biosphere able to withstand insults from toxic materials? To what degree are contaminating substances such as lead and cadmium subject to detoxification by microbiological or chemical action? To what degree are the biological components of the environment able to adapt to change? We know that some bacteria very rapidly assume tolerance to heavy metals. We spent three years showing that nitrogen cycling bacteria in soils from old mining areas were tolerant to very high concentrations of metals, that this tolerance which is genetically controlled can be acquired very quickly, and that the resistant bacteria still do a good job of fixing atmospheric nitrogen or changing its forms in soil.

Some plants can select for metal tolerance, but this, of course, takes much longer. Do all flora and fauna have the ability to adapt and what are the time scales involved? Do some species adapt more readily than others and, if so, why? Is this adaptation controlled by the body size and rate of reproduction? There are something like 10^6 bacteria in one gram of soil and they certainly multiply rapidly. If it is linked to size, there is far less hope for the rhinoceros, cow or human.

I will close on an optimistic note and suggest that we have woken up in time to the damage we are doing to the environment. We are activating laws for the protection of soil, water, the ecosystem and man. However, if we are to react in a timely fashion to the effects of increasing industrial activities and urban development brought about by the continuing population explosion, there is a need for well-focussed research at local, regional and global levels. It is of paramount importance that we address the right problems. Let us not forget that soil is a non-renewable resource. Its protection is essential.

Is there sufficient national and international co-ordination and collaboration? And are there sufficient financial resources being made available? These are the questions I will leave you with.

Note

Reviews of more recent research in this area include Thornton 1991, 1992a, 1992b, 1996a, 1996b, 1997.

References

Barltrop, D., Strehlow, C.D., Thorton, I. & Webb, J.S. (1975) Absorption of lead from dust and soil. *Postgraduate Med. Jnl.*, **51**, 801–804.

Bowie, S.H.U. & Webb, J.S. (ed., 1980) *Environmental Geochemistry and Health.* London: Royal Society.

Brebner, J. (1986) *The Role of Soil Ingestion in the Trace Element Nutrition of Grazing Livestock.* Unpublished PhD thesis. University of London.

Brebner, J., Suttle, M.F. & Thornton, I. (1986) Assessing the availability of ingested soil cobalt for the synthesis of vitamin B_{12} in the ovine rumen. *Proc. Nutritional Society*, **46**, 766A.

Britton, J.W. & Goss, H. (1946) *J. Amer. Vet. Med. Assoc.*, **108**, 176.

Colbourn, P. & Thornton, I. (1978) Lead pollution in agricultural soils in the Southern Peak District. *J. Soil Sci.*, **29**, 513–526.

Cotter-Howells, J. (1991) *Lead Minerals in Soils Contaminated by Mine Waste: Implications For Human Health.* PhD thesis. University of London.

Cotter-Howells, J. & Thornton, I. (1991) Sources and pathways of environmental lead to children in a Derbyshire mining village. *Env. Geochem. Health*, **13**, 127–135.

Culbard, E.B., Thornton, I., Watt, J., Wheatley, M., Moorcroft, S. & Thompson, M. (1988) Metal contamination in British urban dusts and soils. *J. Env. Qual.*, **17**(2), 226–234.

Darnley, A.G., Bjorklund, A., Bolviken, B., Gustavsson, N., Koval, P.V., Plant, J.A., Steenfelt, A., Tauchid, M., Xie, Xuejing (1995). A global geochemical database for environmental resource management. *Earth Sciences 19.* Paris: UNESCO Publishing.

Davies, D.J.A. (1987) An assessment of the exposure of young children to lead in the home environment. In *Lead in the Home Environment*, ed. I. Thornton & E.B. Culbard, pp. 189–196. London: Science Reviews.

Davies, D.J.A. & Thornton, I. (1987) The influence of house age on lead levels in dusts and soils in Brighton, England. *Env. Geochem. Health*, **9**(3/4), 65–67.

Davies, D.J.A., Thornton, I., Watt, J., Culbard, E., Harvey, P., Delves, H., Sherlock, J., Smart, G., Thomas, J.F. & Quinn, M. (1990) Lead intake and blood lead in 2-year-old UK urban children. *Sci. Total Env.*, **90**, 13–29.

Demetriades, A., Ottesen, R.T. & Locutura, J. (1990) Geochemical mapping of western Europe towards the year 2000. *Geological Survey of Norway (NGU) Report 90-105*, nine pages and 10 appendices.

Environmental Protection Agency (1986) *Air Quality Criteria for Lead.* EPA/600/8-83/028 F, June, final draft.

Farago, M.E., Kavanagh, P., Blanks, R., Kelly, J., Kazantzis, G., Thornton, I., Simpson, P.R., Cook, J.M., Delves, H.T. & Hall, G.E.M. (1998) Platinum concentrations in urban road dust and soil, and in blood and urine in the United Kingdom. *Analyst*, **123**, 451–454.

Ferguson, W.S., Lewis, A.H. & Watson, S.J. (1943) The teart pastures of Somerset. I: the cause and case of teartness. *J. Agric Sci.*, **33**, 44.

Fulton, M., Thomson, G., Hunter, R., Raab, G., Laxen, D. & Hepburn, W. (1987) Influence of blood lead on the ability and attainment of children in Edinburgh. *Lancet*, **1**, 1221–1226.

Gilfillan, S.C. (1965) Lead poisoning and the fall of Rome. *J. Occupational Med.*, **7**, 53–60.

Garrels, R.M. & Christ, C.L. (1965). *Solutions, Minerals and Equilibria.* London: Harper and Row.

Healy W.B. (1968) Ingestion of soil by dairy cows. *NZ J. Agric. Res.*, **11**, 487–499.

Hunt, A., Johnson, D., Watt, J., Crutcher, R. & Thornton, I. (1988) A general purpose scheme for classifying lead particles by SEM/EDX. In *Indoor and Ambient Air Quality*, ed. R. Perry & P. Kirk, pp. 539–546. Selper Ltd.

Hunt, A., Johnson, D.L. & Thornton, I. (1991) Descriptive apportionment of lead in house dust by automated SEM. *Wat. Air Soil Pollut.*, **57/58**, 69–77.

Hunt, A., Johnson, D., Watt, J.M. & Thornton, I. (1992) Characterising the sources of particulate lead in house dust by automated scanning electron microscopy. *ES&T*, **26**, 1513–1532.

———— (1993) Apportioning the sources of lead in house dusts in London Borough of Richmond, England. *Sci. Total Env.*, 183–206.

Johnson, D., Watt, J.M. & Culbard, E.B. (1985) Automated image analysis and scanning electron microscopy as an aid to tracing the sources of lead and other dusts. In *Proc. Conf. Heavy Metals in the Environment, Vol. 2*, pp. 404–406. Edinburgh: CEP Consultants.

Kelly, J., Thornton, I. & Simpson, P.R. (1996) Urban geochemistry: a study of the influence of anthropogenic activity of the heavy metal content of soils in traditionally industrial and non-industrial areas of Britain. *Appl. Geochem.*, **11**, 363–370.

Lansdown, R.G., Shepherd, J., Clayton, B.E., Delves, H.T., Graham, P.J. & Turner, W.C. (1974) Blood lead levels, behaviour and intelligence: a population study. *Lancet*, **1**, 538–541.

Lyth, O. (1946) Endemic fluorosis in Kweichow, China. *Lancet.* 233–235.

Leech, A., Howarth, R.J., Thornton, I. & Lewis, G. (1982) The incidence of bovine copper deficiency in England and the Welsh borders — the current situation. *Veterinary Record*, **111**, 203–204.

Lewis, A.H. (1943) The teart pastures of Somerset. II: Relation between soil and teartness. *J. Agric. Sci.*, **33**, 52.

Maskall, J.E. & Thornton, I. (1989) The mineral status of Lake Nakuru National Park, Kenya: A reconnaissance survey. *African J. Ecol.*, **27**, 191–200.

_________ (1991) Trace element geochemistry of soils and plants in Kenyan conservation areas and implications for wildlife nutrition. *Env. Geochem. Health*, **13**, 93–107.

_________ (1992) Geochemistry and wildlife nutrition factors affecting trace element uptake from soils to plants. *Trace Substances in Environmental Health*, XXV, 217–232 (Supplement to Vol. 14 of *Env. Geochem. Health*, Proceedings of a conference held in Columbia, Missouri, May 1991).

Moir, A.M. & Thornton, I. (1990) Lead and cadmium in urban allottments and garden soils and vegetables in the United Kingdom. *Env. Geochem. Health*, **11**, 113–120.

Morgan, H. (1988) Metal contamination at Shipham. *Sci. Total Env.*, **75**, 11–20.

Needleman, H.L., Gunnoe, C., Leviton, A., Reed, R., Peresie, H., Maher, C. & Barrett, P. (1979) Deficits in psychologic and classroom performance of children with elevated dentine lead levels. *New England J. Med.*, **300**, 689–695.

Nriagu, J.O. (1973) Lead orthophosphates. II: Stability of chloropyromorphite at 25°C. *Geochim. Cosmochim. Acta*, **37**, 367–377.

_________ (1974) Lead orthophosphates. IV: Formation and stability in the environment. *Geochim. Cosmochim. Acta.*, **38**, 887–898.

Royal Commission on Environmental Pollution (1983) Lead in the environment. Ninth *Rept CMND 8852*. London: HMSO.

Plant, J.A. & Thornton, I. (1986) Geochemistry and health in the United Kingdom. In *Proc. First Int. Symp. Geochem. Health*, ed. I. Thornton, pp. 1–15. London: Science Reviews Ltd., Northwood, U.K.

Shaper, A.G. (1979) Epidemiology for geochemists. *Philosophical Trans. R. Soc. London*, **B288**, 127–136.

Smith, R.M. & Martell, A.E. (1976) *Critical Stability Constants. Vol. 4: Inorganic Complexes*. New York: Plenum Press.

Strehlow C.D. & Barltrop, D. (1988) The Shipham report: health studies. *Sci. Total Env.*, **75**, 101–133.

Suttle, N.F., Alloway, B.J. & Thornton, I. (1975) An effect of soil ingestion and utilisation of dietary copper by sheep. *J. Agric. Sci.*, **84**, 249–254.

Suttle, N.F. (1981) Predicting the effects molybdenum and sulphur concentrations on the absorbability of copper in grass and forage crops to ruminants. In *Proc. Fourth Int. Symp. Trace Element Metabolism in Man and Animals*, pp. 545–548. Canberra: Australian Academy of Science.

Thornton, I., Kershaw, G.F. & Davies, M.K. (1972) An investigation into copper deficiency in cattle in the Southern Pennines. I: Identification of suspect areas using geochemical reconnaisance followed by blood copper surveys. II: Response to copper supplementation. *J. Agric. Sci.*, **78**, 157–171.

Thornton, I. (1977) Biogeochemical studies on molybdenum in the United Kingdom. In *Molybdenum in the Environment, Vol. 2*, ed. W.R. Chappell and K.K. Petersen, pp. 341–369. Basle: Marcel Dekker.

——— (1980) Geochemical aspects of heavy metal pollution and agriculture in England and Wales. In *Inorganic Pollution and Agriculture*, pp. 105–125. London: H.M.S.O.

——— (1983) Soil-plant-animal interactions in relation to the incidence of trace element disorders in grazing livestock. In *Trace Elements in Animal Production and Veterinary Practice*, ed. M.F. Suttle, R.G. Gunn, W.M. Allen, K.A. Linklater & G. Wiener, pp. 39–49. Edinburgh: British Society of Animal Production.

Thornton, I., Culbard, E.B., Moorcroft, S., Watt, J., Wheatley, M., Thompson, M. & Thomas, J.F.A. (1985b) Metals in urban dusts and soils. *Env. Technol. Lett.*, **6**, 137–144.

Thornton, I. (1987) Mapping of trace elements in relation to human disease. *Clinical Nutrition*, **6**, 97–104.

——— (1988) The Shipham report: metal content of soils and dusts. *Sci. Total Env.*, **75**, 21–39.

——— (1991) Metal contamination of soils in urban areas. In *Soils in the Urban Environment*, ed. P. Bullock & P.J. Gregory pp. 47–74. Oxford: Blackwell Scientific Publications.

——— (1992) Sources and pathways of cadmium in the environment. In *Cadmium in the Human Environment: Toxicity and Carcinogenicity*, eds. G.F. Nordberg, R.F.M. Herber & L. Alessio, pp. 149–162. Lyon: International Agency for Research on Cancer.

_________ (1993) Environmental geochemistry and health in the 1990s: a global perspective. *Appl. Geochem.* (Suppl. issue), **2**, 203–210.

_________ (1995) Sources and pathways of arsenic in south-west England: health implications. In *Exposure and Health Effects of Arsenic*, ed. W.R. Chappell, C.O. Abernathy & C.R. Cothern, pp. 61–70. Northwood, England: Science Reviews Inc.

_________ (1997) Risk assessment related to metals: the role of the geochemist. *Proc. Int. Workshop on Risk Assessment of Metals and their Inorganic Compounds.* Ottawa, Ontario: International Council on Metals and the Environment (in press).

_________ (1996a) Sources and pathways of lead in the environment — implications to human exposure. In *Lead into the Future*, pp. 217–236. London: Institute of Mining and Metallurgy.

_________ (1996b) Impacts of mining on the environment: some local, regional and global issues. *Appl. Geochem.*, **11**, 355–361.

Thomson, I., Thornton, I. & Webb, J.S. (1972) Molybdenum in black shales and the incidence of bovine hypocuprosis. *J. Sci. Food Agric.*, **23**, 871–891.

Underwood, E.J. (1962) Cobalt. In *Trace Elements in Human and Animal Nutrition*, 2nd ed., pp. 123–156. New York and London: Academic Press.

Watt, J.M. & Johnson, D.L. (1987) Characterisation of dusts by scanning electron microscopy with energy dispersive X-ray spectroscopy. In *Lead in the Home Environment*, ed. I. Tjornton and E.B. Culbard, pp. 85–96. Northwood: Science Reviews.

Webb, J.S., Nichol, I., Foster, R., Lowenstein, P.L. & Howarth, R.J. (1973) *Provisional Geochemical Atlas of Northern Ireland.* London: Applied Geochemistry Research Group, Imperial College of Science and Technology.

Webb, J.S., Thornton, I., Howarth, R.J., Thompson, M. & Lowenstein, P.L. (1978) *The Wolfson Geochemical Atlas of England and Wales.* Oxford University Press.

Zheng, B. & Hong, Y. (1988) Geochemical environment related to human endemic fluorosis in China. In *Geochemistry and Health*, ed. I. Thornton, pp. 93–96. Northwood: Science Reviews Ltd.

CHAPTER 5

VIRTUAL ECOLOGY

JOHN D. WOODS*

Introduction

My inaugural lecture will address one of the major problems in oceanography: understanding the geographical distribution of life in the top kilometre of the ocean in terms of the laws of marine physics, chemistry and biology. The subject is inter-disciplinary, and to make progress it is necessary to bring together skills in science and engineering, in theory and experiment, and in the new technology of computer simulation.

That new technology promises to revitalise the subject of biological oceanography, which has made much less progress than physical oceanography in the last 100 years. I shall describe the new world of virtual ecosystems, which contain information that cannot be obtained directly from observations. This leads to the science of virtual ecology, which will change the way we learn how life in the sea varies both geographically and in time: diurnally, seasonally and inter-annually, as well as over decades and centuries as climate changes. It has taken ten years to develop this new technology. It is now ready to address the fundamental issues of biological oceanography.

In my lecture, I shall show what it can do to address a number of classical phenomena. Then I shall give a hint of its potential by describing

*Professor of Oceanography, University of London, Imperial College of Science, Technology and Medicine.

a virtual ecosystem in which the plankton adapt by mutation and natural selection, and another in which they influence global warming. The lecture will close with a glimpse of the future of virtual ecology applied to a wide range of practical problems in the ocean, including biodiversity and the impact of ozone depletion, marine epidemiology and fisheries.

The Discovery of Laminar Flow in the Ocean

Biological oceanography is a synthesis of marine physics, chemistry and biology. I came into the subject through physics. When I was a research student in the old Huxley Building 30 years ago, I made a discovery which later turned out to be the key to virtual ecology.

My discovery concerned the nature of flow in the ocean. In the 1960s, it was believed to be universally and continuously turbulent. That was consistent with the fact that ocean currents have a very high Reynolds Number, orders of magnitude greater than the critical value for transition from laminar to turbulent flow. It was known that the turbulent diffusivity varied with depth in response to density gradient, following Richardson's Law, but it was assumed that turbulence prevailed everywhere. That paradigm had been established by the greatest authority on turbulence: Sir Geoffrey Taylor. It appeared in all the textbooks. And it underpinned biological oceanography, for which changes in turbulent intensity played a central role in theories of the annual cycle of life in the sea by Sverdrup and others.

In the 1960s, I discovered that the paradigm was wrong. In fact, oceanic flow is naturally laminar, with turbulence occurring only locally in transient bursts. The analysis by G.I. Taylor had been performed on measurements from an anomalous region, the Kattegatt, which provided a misleading model for the world ocean. My discovery was made in the classical way — by flow visualization. Using diving apparatus, I filmed the patterns made by dye injected into the water as it was dispersed by currents and modulated by internal waves. I shall now show a film that I made in 1965 (Fig. 1).

Fig. 1. The discovery that the flow is predominantly laminar in the seasonal thermocline has played a key role in developing an understanding of the plankton ecosystem. This photograph (from J. Woods, *J. Fluid Mech.*, **32**, 791, 1968) shows dye injected some hours earlier into the seasonal thermocline. The streaky appearance indicates laminar flow. The undulations reveal an internal wave.

Video of laminar flow

The film shows a thin sheet of dye in the seasonal thermocline. The primary motion is obviously laminar. The dye is not universally dispersed by small-scale eddying.

The dye sheet undulates in response to the passage of an internal wave with a wavelength of ten metres and a period of five minutes. As the wave goes past, the fluid motion associated with its rise and fall produces a vertical shear of the horizontal flow, with peak value on the crest of the wave. I was able to show that this shear is sufficient to cause Kelvin-Helmholtz instability, which leads to ripples that grow to billows in about one minute. The Reynolds number of each billow is less than 100, but it just exceeds Peter Bradshaw's criterion for breakdown into a cascade of smaller turbulent eddies, as we see in the film. This work was done in the Mediterranean Sea, where conditions are relatively calm in summer, but I later repeated it in the Gulf Stream between Florida and the Bahamas, where conditions are quite the opposite. The result was the same there (in fact, the main source of turbulence came from the tails of curious sharks).

The results of these experiments now appear in all the textbooks as the standard paradigm for flow in the ocean. Apart from a turbulent surface layer powered by the wind, the flow is normally laminar with local, short-lived bursts of turbulence at low Reynolds number occupying about 5% of the volume at any instant. A planktonic organism may encounter such a burst of turbulence only a few times in its life: about the same probability that a resident of Florida would experience a hurricane.

In 1966, I showed this film at a Geophysical Fluid Dynamics seminar in Cambridge. During the questions, an elderly gentleman in the front row asked me if I would kindly define my terms: What exactly did I mean by the words "turbulence" and "laminar flow"? I did not know my questioner, and only afterwards discovered that I had just defined turbulence for Sir Geoffrey Taylor. A happy start to a research career in oceanography.

My discovery has proved to be quite important in many aspects of oceanography, from submarine warfare to climate. The subtle interplay between laminar and turbulent flows described by the new paradigm helps explain many previously puzzling phenomena in marine ecology. Which brings me to the main theme of my lecture: the ecosystem in the upper ocean.

The Scientific Problem

During the last ten years, I have shifted my research focus from physical oceanography to biological oceanography. The two subjects are closely linked in nature, but the practice of physical and biological oceanographers is very different. During the last century, there has been enormous progress in understanding the physics of the ocean — the waves, currents, turbulence and density distribution. New technology, discoveries, theories and experiments have all contributed to produce improved understanding.

Biological oceanography, on the other hand, has not enjoyed an equivalent revolution. The subject in the 1990s has not changed much since the 1890s, as Eric Mills's (1989) perceptive history of the subject has recently shown. Biological oceanography has two aims. The first is to map the geographical distribution of life in the sea, allowing for its seasonal and inter-annual variation. The second is to explain those distributions in terms of the laws of marine physics, chemistry and biology.

Why has there been so little progress in the last 100 years, given the rapid strides taken by physical oceanography? And why has biological oceanography, which is a branch of marine ecology, not benefitted from the advances in terrestrial ecology? A recent discussion meeting at the Royal Society sought to explain the lack of interaction between marine and terrestrial ecology, but failed. That set me thinking. Viewed from the perspective of a physical oceanographer, why is biological oceanography so isolated and stagnant? The distinguished Scottish fisheries oceanographer, John Steele, argues that it is because the physical environment plays so much greater a role in the ocean ecosystem. That is true, but it is not a sufficient explanation.

I believe that I have identified the real explanation. If I am right, we can understand why progress has been so slow, and we must conclude that there is little hope of accelerating it, at least by the traditional methods.

Analysis of the Problem

The reason can be found from information theory. I have recently calculated how much information is needed, in principle, to address the mission of biological oceanography. Consider a water column one square metre in horizontal area and one kilometre deep. How much information would be banked if we were to describe the ecosystem in that water column over one year?

Plankton

It contains of order one billion living organisms in a wide variety of species, which we can group into a limited number of functional groups or guilds following normal practice in theoretical ecology. The living organisms are mainly microscopic plankton. They drift around in the ocean, supported against gravity by the Archimedes force. Unlike plants and animals on land, they do not need to invest a significant fraction of their resources into structures that can resist gravity and storms. There are no trees or giraffes in the sea. (Kipling would have written very different "just so" stories for animals in the ocean.) Although tiny, plankton are very much more productive per unit biomass than their terrestrial equivalents; and they fix more carbon each year.

Most plankton can swim fast enough to change their ambient environment significantly. This is provided they move vertically. However, they cannot change their ambient environment by swimming horizontally as the correlation scale of the horizontal variation of environment fields is 100 times greater than in the vertical, for reasons that are well understood in terms of geophysical fluid dynamics. This means that plankton behaviour can be treated as a one-dimensional scientific problem.

Geographical variation in the plankton is essentially a one-dimensional response to changing climatological conditions as the water column is carried around the ocean by the currents. The ecology of plankton is very different from the ecology of fish which *can* swim great distances. But fish are rare and do not yet feature in biological oceanography.

Environment

The environment in the water column has physical and chemical variables. The physical variables are velocity, temperature, salinity, light, turbidity and *turbulence*. The chemical variables are carbon, nitrogen, silica, phosphate in a number of chemical forms, in particles or dissolved in the sea water.

Information content

I estimate that it would take some trillion four-byte words of data to describe the diurnal and seasonal variation in these physical, chemical and biological variables in the water column. With that information, it would be possible to describe the growth and behaviour of every plankter living in the water column during the annual cycle: its response to ambient environment; its influence on the environment; its interaction with other plankters; its reproduction, natural death, or premature consumption; and its life expectancy and other demographic properties. With that information, it would be possible to study the population dynamics of the plankton and their influence on the environment, and to understand the distribution of life in the ocean. Unfortunately, we do not have such data.

Information content of experimental data

I have recently computed the information content of recent biological oceanography research cruises. The most elaborate international collaborative projects, like those of the Joint Global Ocean Fluxes Study

(JGOFS), yield less than one million words of data. That is less than one millionth of the data needed to achieve the mission of biological oceanography. There is no prospect of significantly increasing that fraction. I doubt whether, even in his most optimistic moments, Professor Krebs (Chief Executive Officer of NERC) expects to be able to fund a millionfold increase in observations for biological oceanography.

There are two direct results of such massive undersampling:

1. Maps purporting to display the distribution of life in the ocean should be treated with caution as they are likely to be aliased.
2. Even if that problem could be overcome, there is no prospect of understanding the distributions. This is because the empirical basis for population dynamics is inadequate. Unlike bird watchers, biological oceanographers cannot ring individual plankters and track them as they interact with their mates, predators and prey. There are no data for demographic calculations: no repeated censuses; no parish registers.

Consequences

There are important practical consequences of these deficiencies. The show-stopping inability to make enough of the right kind of observations means that biological oceanographers cannot provide reliable numbers for important global and regional phenomena controlled by the marine ecosystem:

1. Global warming is influenced by the changing carbon uptake of the ocean, largely through biological production.
2. The capacity of the ocean to process contaminants and so maintain the health of the ocean is controlled by the plankton.
3. There are important military and other applications for which it is desirable to predict the changing activity of plankton.

Thus, we do not have the option of putting biological oceanography into the "too difficult" category and turning our efforts onto the more tractable scientific problems of marine physics, chemistry and biology, for which it *is* possible to make rapid progress.

Solution

The challenge is to accelerate progress in biological oceanography, which seeks to synthesise those basic disciplines. I believe that I have found a way to achieve that. My solution to our inability to observe and analyse the natural ecosystem is to create a *surrogate* that we can simulate and analyse. I call that surrogate a *virtual ecosystem*. It is created by integrating a mathematical model comprising a set of equations and parameters which describe the basic processes of marine physics, chemistry and biology.

Mathematical modelling of the ocean ecosystem is not new, but this is the first attempt to create a realistic virtual ecosystem in which all the principal processes are simulated at the level of individual plankters, and at the level of their communities, so that it realistically describes the dynamics of the plankton populations. Using new modelling technology, it has proved possible to simulate the main phenomena of the natural ecosystem with an information density that approaches the target of a terraword/m^2y. In fact, within a decade or so, increased computer power will make it possible to create simulations at the target information density.

Technology

The secret of creating virtual ecosystems lies in a new method of integrating the mathematical equations. I call this new technology the *Lagrangian Ensemble* (LE) method. The growth and behaviour of each plankter in a water column is simulated by integrating the biological equations separately for it. The individual plankters develop and behave differently because they follow different trajectories up and down through the water column, experiencing different variation of ambient environment.

The model integration proceeds in half-hour time steps. At each time step, the LE method computes the changed position and biological state of every plankter, and the ensemble statistics of the plankter populations in each one metre layer of the environmental description. The simulation is started with prescribed initial conditions taken from climatology. It is then driven by surface fluxes of sunlight, heat, water and gases.

The LE method is explicit. It is, therefore, quite different from the Monte Carlo technique used, for example, in fisheries models. It is, to use an expression familiar to meteorologists, a primitive equation model. It, therefore, has the potential to make forecasts given suitable data. Meanwhile, it can improve understanding through simulation based on climatological forcing.

Demonstration

I shall now illustrate the LE method by a virtual ecosystem designed to simulate plankton behaviour in the North Atlantic Ocean off the Azores (Fig. 2). I shall show three aspects:

1. The properties of a single planktonic plant.
2. Statistics of the population of plants.
3. Behaviour of the zooplankton which eat the plants.

Audit trail of a diatom over five days in April

The core of the LE method is its simulation of the growth and behaviour of each plankter separately. When we analyse the data set that is the virtual ecosystem, we can extract audit trails of individual plankters (Fig. 3). Here, we see the development of a diatom over five days in April at a site in the Atlantic Ocean off the Azores in four slides. In the first, we note the trajectory of the plankton as it is displaced randomly up and down in the surface turbulent layer, and later sinks slowly through the laminar flow regime below. The depth of the mixing layer varies diurnally, being driven down at night by convection which is quenched by solar heating during the day. (Even quite small changes in radiation affect the depth of the surface mixing layer, as we shall see later.) Next, we see the variation of ambient irradiance in the photosynthetic waveband (400–700 nm) experienced by the diatom, and

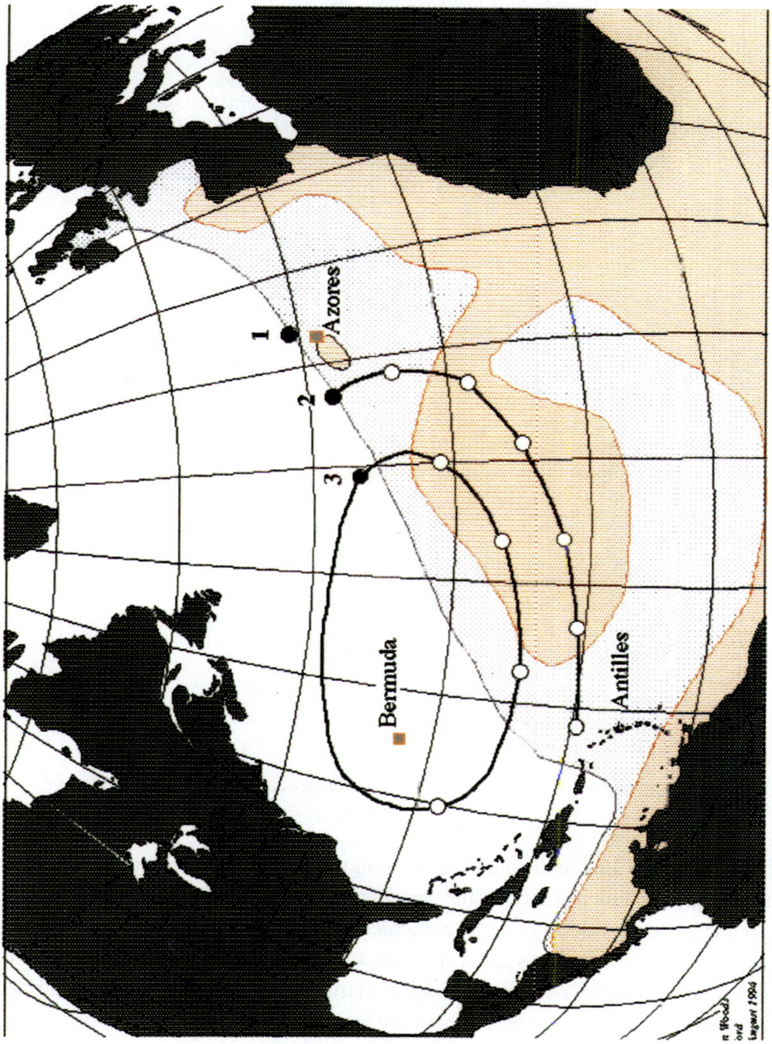

Fig. 2. Tracks followed by water columns drifting with the circulation in the North Atlantic. The white circles indicate 1 January each year.

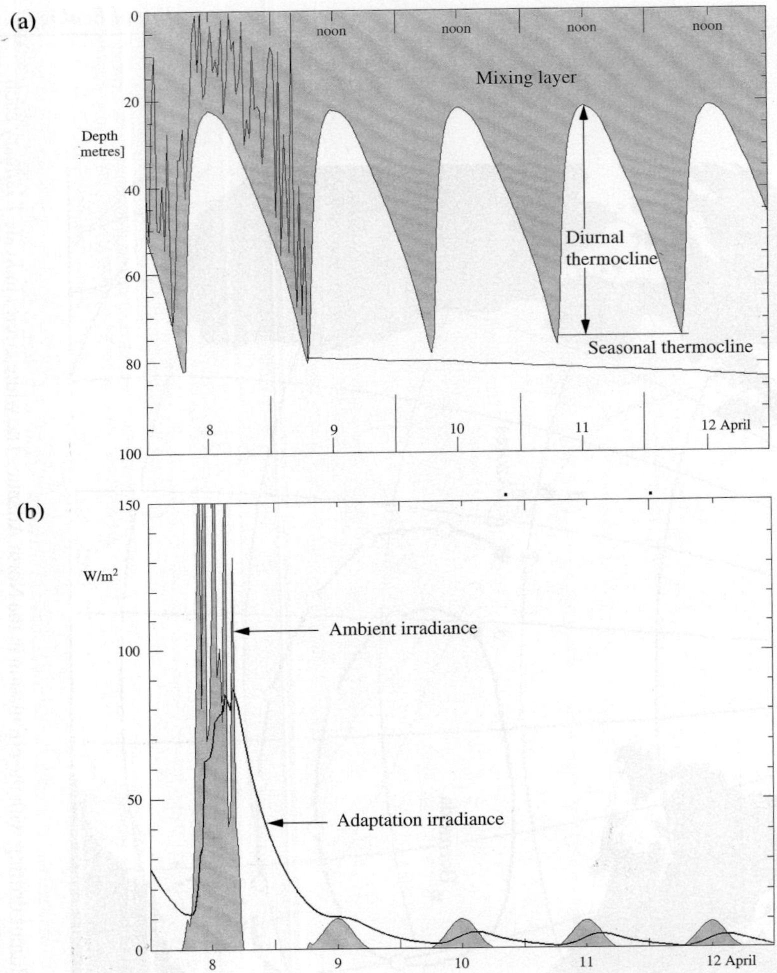

Fig. 3. The Lagrangian Ensemble method of integrating a model of the upper ocean ecosystem is unique because it simulates the growth and behaviour of individual planktonic organisms, while accurately describing the dynamics of their populations in the context of a realistic environment. This five-day audit trail for a single phytoplanker (at location 1 in Fig. 2) shows: (a) the plankter's variation of depth, starting initially in the turbulent surface layer, but then being subducted into the laminar flow regime below; (b) the ambient irradiance (400–700 nm) experienced by the plankter and the irradiance to which it is adapted; (c) the rate of energy uptake by photosynthesis: the horizontal line shows the plankter's energy loss by respiration; and (d) the energy stored in the plankter; when it exceeds 145 μJ, the cell divides.

(c)

(d)

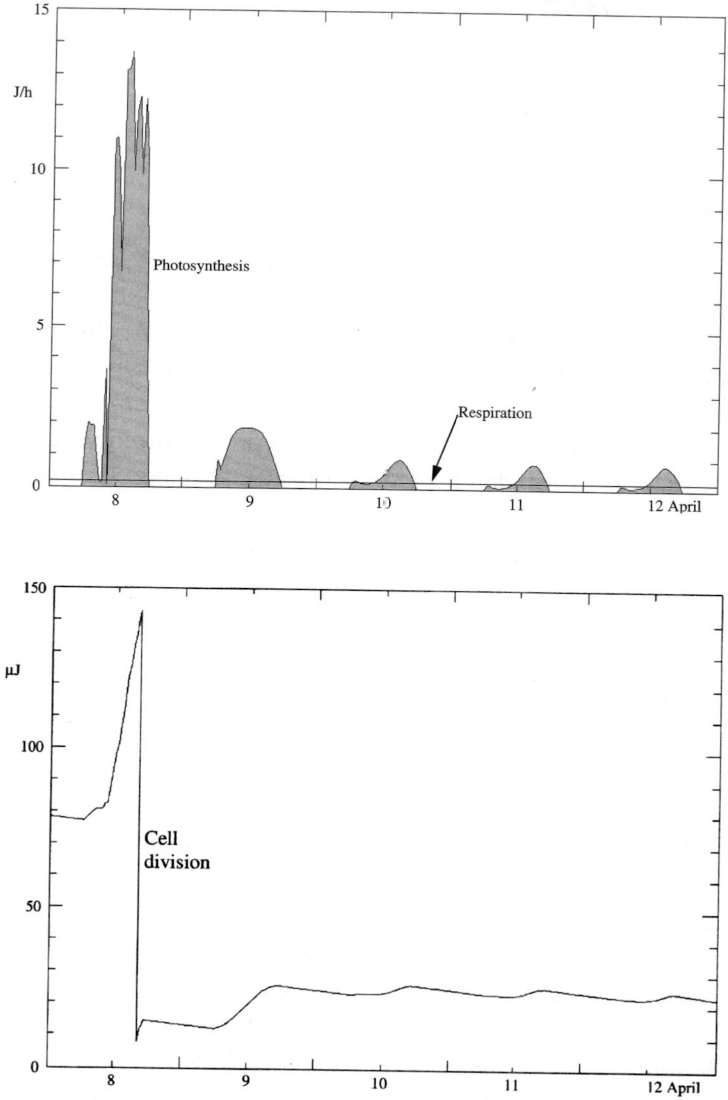

Fig. 3. (*Continued*).

the way it photo-adapts to the recent light level. The third figure shows the rate of energy uptake by photosynthesis, compared with the rate of energy consumption, or respiration. The final figure in this set shows the energy store in the diatom. When it exceeds a critical value of 145 μJ, the cell divides.

Demographic variables of diatoms

The developments in every plankter are recorded in a parish register, which is later analysed to provide demographic data about the population (Fig. 4). Here, we look at the time series of four demographic properties of the diatom population. The first shows how the total number of diatom cells varies throughout the year. The maximum in spring accords well with observations, and is called the spring bloom. Its timing is controlled by the reduction in the depth of the surface turbulence later in spring. Its end is controlled by two factors: a reduction in the birth rate when the nutrients run out; and an increase in grazing rate as the zooplankton grow. I shall have more to say about both processes shortly. The second figure in this set shows how the life expectancy of the diatoms varies throughout the year. The third and fourth figures show the annual variation of birth and death rates.

Diurnal migration of copepods

In order to understand the complex processes that are responsible for these demographic changes, we need to examine the vertical distribution of the environment and the plankton population living in it (Fig. 5). We start with the most important property: turbulence. The photograph reminds us that the flow is mainly laminar below the surface mixed layer, which varies in depth from a minimum at the summer solstice to a maximum at the end of the cooling season, which normally occurs near the spring equinox. This asymmetry in the seasonal cycle of mixed layer depth (three months of shoaling and nine months of deepening) accurately simulates observations.

Next, we examine the depth distribution of the phytoplankton, which are concentrated in the top 50 metres where there is normally sufficient light for photosynthesis. I say "normally" because in spring, the concentration of diatoms becomes so great (billions per square metre) that the light cannot penetrate them and photosynthesis stops until the zooplankton have thinned them out (Fig. 5b). As the zooplankton graze on this rich supply of food, they grow rapidly, reach maturity and start to grow eggs which are hatched while there is still plenty of food available. The new generation of herbivorous zooplankton is released, increasing the population density by two orders of magnitude. They are preyed upon by carnivorous zooplankton, many of which hunt visually. Other carnivores hunt acoustically, and others passively net their prey. In the model used to generate *this* virtual ecosystem, all carnivores hunt visually. Their success depends on the visibility of their prey, which are invisible on a dark night, or deep in the ocean. So it pays to swim down during the day and rise up to eat at night. Larger prey are more visible than smaller ones. Hence the newly hatched larvae are less vulnerable, which is just as well because they cannot swim fast enough to get down into the dark depths each day.

Observations show how zooplankton descend during the day, presumably to avoid detection, and rise up at night to feed. The slide shows measurements made by colleagues from the Southampton Oceanography Centre led by Howard Roe. The acoustic signature comes from the larger zooplankton which produces the strongest echoes: they are the bigger carnivores chasing the herbivores, and avoiding being detected by those that prey on them.

This diurnal descent of the zooplankton is accurately simulated in the virtual ecosystem, as we see in the next 14 slides, showing the vertical distribution at hourly intervals. Note that the extent of the migration is less in spring, when the population is dominated by slow-swimming larvae. It picks up in summer as they grow bigger and stronger — and more visible to the carnivores preying on them. Even the adults do not have to migrate down in spring when the phytoplankton are so prolific that they block out the sun. The herbivores can then stay undetected in the top 50 metres, feeding day and night, when the food supply is at its best. This helps to explain recent observation of the rapid response of zooplankton to an artificially stimulated bloom.

Geographical Variation

Having seen something of the internal working of a virtual ecosystem, we now turn to the fundamental challenge of biological oceanography: to map and explain the distribution of life in the sea. Because the behaviour of plankton is one-dimensional, we can address the problem of geographical variation by studying the changes occurring in a virtual ecosystem contained within a water column that is advected around the ocean by the large-scale currents. We select tracks from the velocity distribution simulated by a circulation model of the ocean. The results I presented so far were for a water column off the Azores. Now, I shall show changes that occur over six years in a water column drifting from the Azores to the Antilles (Track 2 in Fig. 2).

Transition to oligotrophy

We start by examining the change in depth of the surface turbulent layer, which is the key to plankton growth (Fig. 5a). We note that the annual maximum depth of the mixed layer is about three times the depth of the euphotic zone (in which phytoplankton grow) off the Azores, but as the water column drifts into the tropics, the mixing penetrates less each winter. The temperature rises each year. So does the solar irradiance. But after a few years, the phytoplankton stop reproducing. Why is that so? Previously, we learnt that the conditions for phytoplankton multiplication is light and shallow mixing. There must be another factor controlling growth. It is the supply of nitrogen, which is the limiting nutrient in this region. Cell division can only occur if the phytoplankter has stored sufficient energy *and* sufficient nitrogen. As the water column enters the tropics, it becomes short of nitrogen. It is entering what biological oceanographers call the *oligotrophic zone*, where observations show low nitrogen, low phytoplankton concentration and low primary production (that is, low rate of phytoplankton growth).

The changes occurring in the virtual ecosystem, as it drifts towards the Antilles, accurately simulate those observations and describe in unprecedented detail the internal changes occurring as the depth of the surface turbulent layer decreases year by year. Nitrogen is available to the phytoplankton in two forms: nitrate and ammonium. The former is injected into the model at the start of the integration and has no other source; the latter is generated internally by zooplankton excretion and microbial action on dead plankton and faecal pellets. The next pair of figures show the uptake and concentration of nitrate in the water (Fig. 5c). Each winter, deep mixing charges the euphotic zone with nitrate, which is consumed by the plankton growth during the spring bloom. As the water column drifts into the tropics, winter mixing decreases and the annual supply of nitrate decreases. The same is true (next pair of slides) for ammonium, with the added complication that some of the nitrogen taken up by plankton is returned as ammonium for the following year's growth. However, that supply also weakens as the water column drifts towards the Antilles. That is why the growth rate declines, despite the plentiful supply of sunlight. The death rates of the phytoplankton by energy starvation and grazing decrease as the population densities of both phytoplankton and zooplankton decline.

These changes are summed up in the population dynamics figure (Fig. 4), which shows the changes in the populations of the phytoplankton and zooplankton, and in the herbivorous and carnivorous predation rates. In fact, as the water mass approaches the Antilles, the herbivores find so little food that they indulge in an orgy of cannibalism before becoming extinct.

Recovery from oligotrophy

Now we turn to Track 3 in Fig. 2 which starts a little to the West and circulates around Bermuda, taking five years to return to its starting point. This passes through the oligotrophic zone for a couple of years,

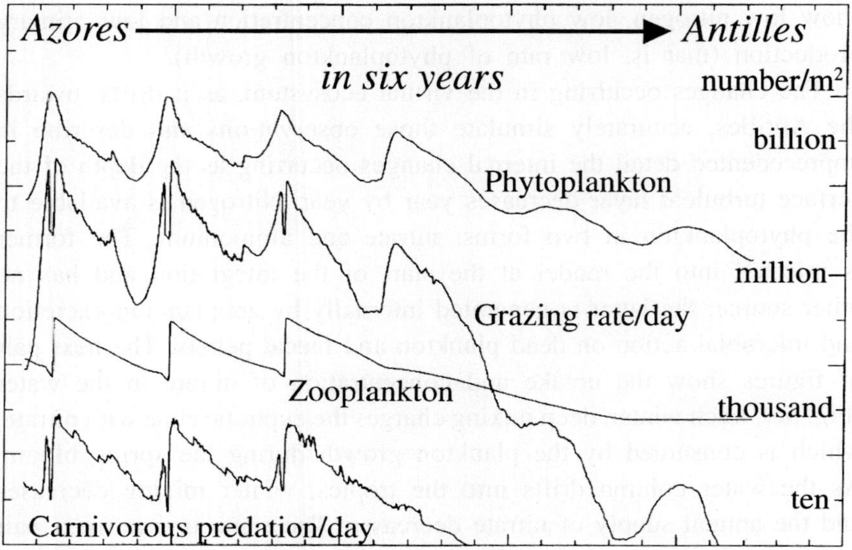

Fig. 4. Variation of plankton population dynamics along Track 2 (see Fig. 2) showing the decline in the phytoplankton population as the water column enters the tropics, and the consequent collapse of grazing rate. The herbivorous zooplankton manage to reproduce each spring in the first three years (note the sharp rise in the number of animals), but fail to reproduce thereafter because of inadequate food supply. The rate at which carnivorous zooplankton eat these herbivores also collapses dramatically in the tropics after reproduction has stopped.

when the nitrate and ammonium supplies become low, but production continues at a level that does not lead to zooplankton extinction. The populations recover as soon as the water column returns to Northern waters, where cooling in winter drives deep convection, bringing up nutrients for the following year's growing season.

Inter-Annual Variation

One of the questions that spring to mind is: how reproducible are the changes that occur as the water column circulates around the Sargasso

Sea? One way to answer that question is to allow the water column to make several laps around the closed circuit and compare the populations in successive laps. We see, by overlapping the five-year time series of demographic variables, that the population dynamics are remarkably repeatable, at least to first order. Why is that so? The model which creates the virtual ecosystem is driven by the fluxes of heat and energy through the sea surface. They are prescribed to follow the regular annual cycle of the climate at each location around the circuit. Thus, the forcing is exactly reproducible. It controls the physical environment and, in particular, the depth of the turbulent layer. The plankton dynamics are controlled overwhelmingly by the turbulence.

That is true to first order. When we examine the differences between the populations along each lap in finer detail, we begin to notice some jitter. This inter-lap variability is greater in the zooplankton population than that in the phytoplankton. That is not entirely unexpected, although this is the first time it has been measured (albeit in a virtual ecosystem). The growth of inter-lap variation with trophic level suggests it is due to predator-prey interaction. The numerical experiment opens the door to an attack on the general problem of predator-prey interactions in the virtual ecosystem.

Population Adaptation by Mutation and Natural Selection

The idea of a virtual ecosystem circulating repeatedly around the Sargasso sea without mixing with its neighbours is, of course, artificial, but it is helpful for clearing the mind about processes occurring in the ocean. More realistic simulations will have to take account of factors neglected in this simulation, notably vertical shear and lateral mixing. However, I believe we can use the simplified model to explore a number of important processes. For example, we can study the consequences of allowing the plankton to adapt to their environment by mutation and natural selection. I would like to briefly report some preliminary results of an ongoing investigation of that process.

This study concentrates on the mutation of the energy reserve parameter which we encountered earlier when examining the audit trail of a diatom. There is no experimentally determined value for this parameter. Increasing it leads to a lower birth rate compensated for by a lower rate of infant mortality. Cell division tends to occur in late afternoon, after a day's photosynthesis. If the energy reserve is too low, nocturnal respiration may kill the two new cells before the sun rises the next day. In the numerical experiment, we start with the parameter as being equal to zero (high birth rate, high infant mortality), but allow the value to mutate incrementally at a prescribed mean frequency of one in ten cell divisions. This is an artificially high mutation rate, but I wanted to see whether it was possible to get any significant adaptation before proceeding to more realistic conditions.

We saw earlier that the annual death rate among diatoms is remarkably high. For every 100 families alive at the start of a year, only one will survive to the next year. We expect natural selection to favour those with values of the energy reserve parameter that is best fitted to the environment. We run the model for 30 years, during which time the water mass laps the circuit six times. And we examine the distribution of the energy reserve parameter among the diatom population at the end of each year. The results, shown in the next pair of slides, can be summarised as follows (Fig. 5). The initial conditions were $\delta E = 0$. The first lap of five years was a period of transition in which the Gaussian distribution of the parameter was established, with a mean close to the energy consumption by 12 hours of respiration. That is the minimum reserve needed by a diatom which divides just at nightfall if it is not to die during the first night. The distribution has a standard deviation of about 10 μJ. This distribution is stable from lap 2 to lap 6, with minor variation between laps.

These rather provocative results require further investigation before one could justify a research cruise to check them out. But we can conclude that the virtual ecosystem is sufficiently stable to perform such numerical experiments.

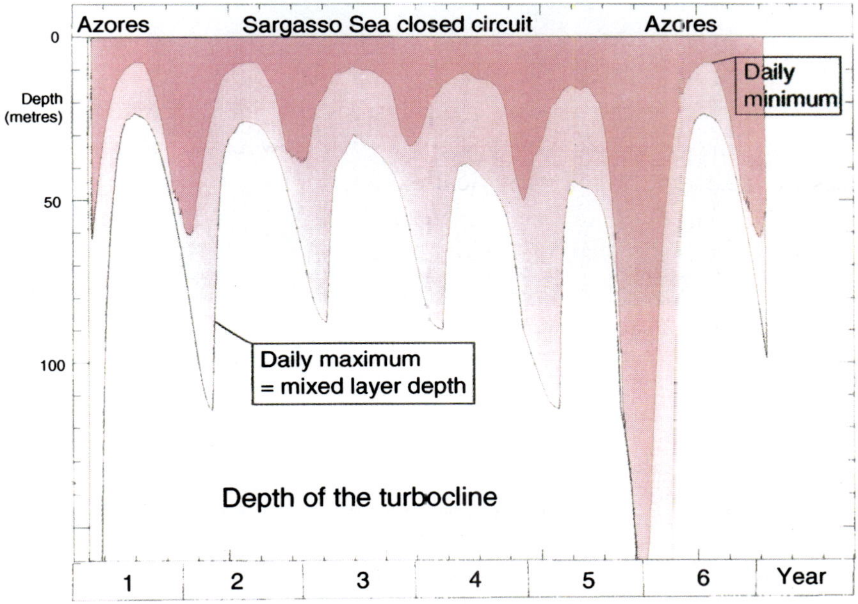

(a)

Fig. 5. Ecological changes occuring in a water column during the five years that it takes to circulate the Sargasso Sea along Track 3 (see Fig. 2). (a) The daily minimum and maximum depths of the surface turbulent layer. Note the reduced depth in the tropics, where the ocean gains more heat each year from the sun than it loses to the atmosphere; (b) The optical environment shown by 400–700 nm isolumes. Note the seasonal undulation and the changes due to enhanced turbidity during the spring bloom; (c) The depth of the surface turbulent layer, showing the relatively shallow winter mixing in the tropic, which controls the ecosystem; (d) Nitrate concentration. Note the annual replenishment each winter when the mixed layer (thin red line) is deepest. The annual maximum depth of the mixed layer is much less in the tropics, leading to a smaller annual injection of nutrients and consequent poor phytoplankton growth; (e) Phytoplankton growth rate; (f) Zooplankton concentration.

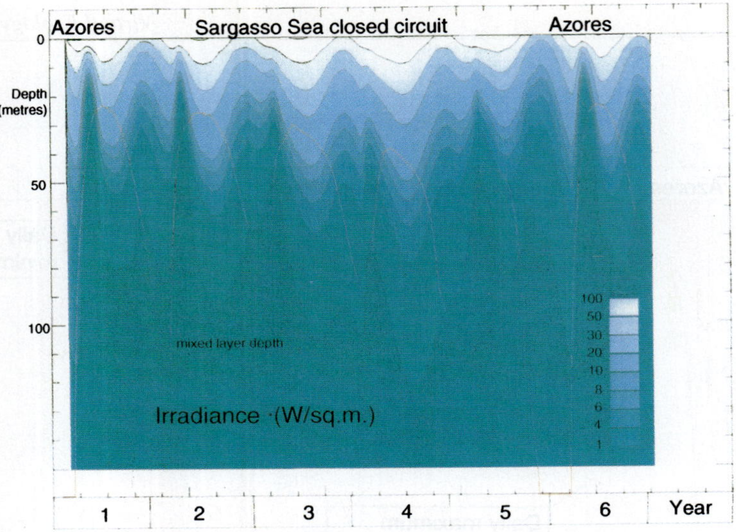

(b)

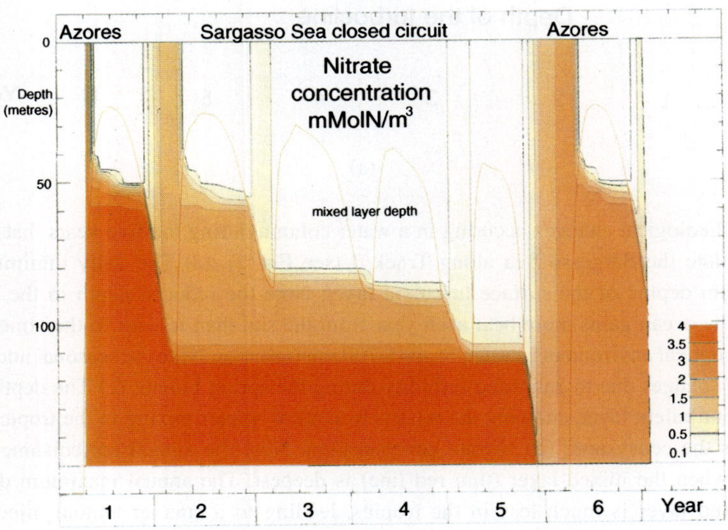

(c)

Fig. 5. (*Continued*).

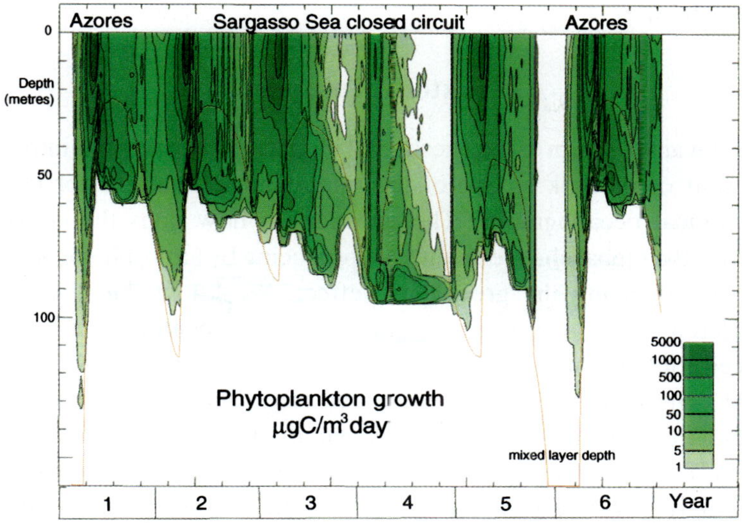

(d)

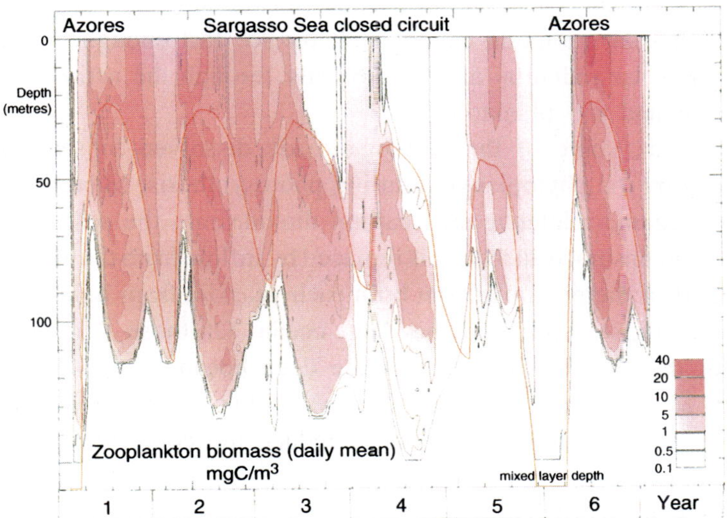

(e)

Fig. 5. (*Continued*).

Plankton Multiplier

Finally, I want to return to the question of climate change which I introduced at the start of my talk as one of the important motivations for research on biological oceanography. The question is: how does the ecosystem respond to the global change that we are producing by atmospheric pollution, which is enhancing the greenhouse effect. We know that the last ice age ended about 15,000 years ago in a massive greenhouse effect which produced global warming. The cause of that rise in atmospheric carbon dioxide concentration is not known for sure. My hypothesis is that it was due to a change in the turbulence in the ocean, which reduced the global carbon uptake by phytoplankton. We have seen that the turbulence is extremely sensitive to the solar or infrared radiation entering the sea. We also know that the ice ages are driven by periodic changes of a few Watt/m^2 in the solar radiation entering the ocean, due to orbital changes in the Earth around the Sun under the gravitational pull of Mars and Venus. Twenty years ago, Sir John Mason showed that such small changes in solar radiation cannot melt the ice fast enough. My theory is that the solar radiation did so indirectly through the action of ocean turbulence and the plankton.

Each year, some of the carbon dioxide dissolved in sea water is consumed by the phytoplankton during photosynthesis. Microbial action on dead organic matter returns carbon dioxide into the sea water. The net amount consumed each year is replaced by inflow from the atmosphere. About half of the phytoplankton born each year die from light starvation and their corpses sink into the deep ocean. Of those that are eaten, much of the carbon also sinks to the deep ocean as fæcal pellets or as dead zooplankton. The total downward flux of carbon carried into the deep ocean in this way helps to reduce the concentration of CO_2 in the atmosphere.

The scientific problem is to determine how the upper ocean ecosystem, responds to a change in the carbon dioxide concentration of the atmosphere, and to discover whether the increasing greenhouse effect will enhance or diminish the oceanic uptake of carbon. To address that problem, I recently carried out an experiment in which the surface radiation flux was progressively increased, in accord with expected trends over the next 50 years. The results were broadly as expected from the following theory.

The mechanism is a subtle one, depending on the effect of infrared (IR) radiation on turbulence in the upper ocean. The greenhouse effect acts as an IR blanket on the ocean, reducing the IR cooling to the atmosphere and, therefore, the depth of buoyant convection, which partially powers the turbulence. As the atmospheric CO_2 concentration rises, the IR blanket thickens, convection reduces and the mixed layer gets shallower (Fig. 6). As we have seen, a shallower mixed layer leads to less nitrogen uptake for the growing season and, therefore, less phytoplankton growth and a smaller carbon flux into the deep ocean.

To summarise, virtual ecosystems *reduce* their uptake of carbon as the atmospheric CO_2 concentration rises. The uptake will be reduced by 10% over the next 50 years as the atmospheric carbon dioxide concentration doubles. This represents *positive feedback* in the greenhouse effect. The plankton multiplier accelerates global warming by the greenhouse effect. This result also supports my theory that the rapid warming at the end of the last ice age was due to the action of the changes of solar radiation on the turbulence.

Model Verification

By now, I hope it has become clear that virtual ecosystems can provide insight into a wide variety of problems in biological oceanography. Before closing, I must say a few words about their credibility. Many scientists are rightly sceptical about computer simulation because, they argue, it is all unconstrained fantasy. As Pyotr Kapitsa remarked: "Theory is all very well, but a great experiment lives for ever". The question is: what constitutes a great experiment in biological oceanography? It should measure a phenomenon that has been simulated in a virtual ecosystem, and so test the model used to create the virtual ecosystem. So far, there have been very few great experiments in biological oceanography; largely

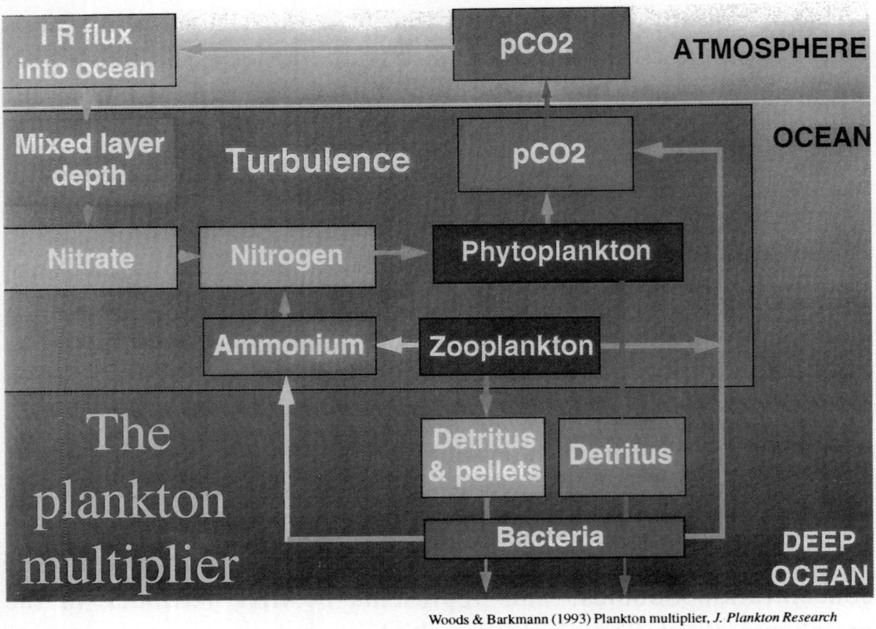

Fig. 6. The plankter multiplier which influences the concentration of carbon dioxide in the atmosphere. When the atmospheric CO_2 concentration rises, it generates an enhanced greenhouse effect which warms the surface of the ocean by IR radiation. The immediate consequence is a reduction in the depth of the turbulent surface layer. That leads to a reduction in nitrogen supply to the euphotic zone and, therefore, less annual phytoplankon growth. This cuts the demand for carbon dioxide, leaving more in the atmosphere. This plankton multiplier is a biological positive feedback which will accelerate global warming in the twenty-first century.

because it is so difficult to collect enough of the right kind of data. During the exploratory phase of a new science, the available observing resources may be dispersed to catch new phenomena: as it matures into an experimental science, they must be concentrated onto describing phenomena predicted by theory.

The credibility of virtual ecology remains in doubt because we have few, if any, great experiments to test its models. Meanwhile, do we have any reason to believe the phenomena simulated in virtual ecosystems? Experience with primitive equation models in meteorology tells us that the simulations are realistic (though not necessarily correct in detail) if the basic equations are sound. That works for models based on equations in physics. How well does it work for ecological models with biological equations? Do we know the biological equations well enough to be sure that they will lead to realistic simulations when incorporated into a primitive equation model?

The nature of laws in physics and biology

This raises an interesting point about the difference between physics and biology. Physics was established at the Big Bang and, according to Einstein, has not changed since then; our task is to discover its laws and the equations and parameters that best describe them. Life was created later ("On the fifth day, God created every living creature"). Biology, too, has laws to be discovered which can be described mathematically. But the biological equations and, in particular, the parameters are not all fixed once and for all. Many are continually evolving; some quite rapidly. The task of marine biologists is to discover the laws by laboratory experiment and to provide first guess values for the parameters. Our task as ecologists is to include in the model the rules for the parameters to adapt by mutation and natural selection, as they do in nature.

Ecological models are primitive equation models with the physical parameters determined by experiment and the biological parameters free to adapt by mutation and natural selection. Given that strategy, I expect ecological models to behave as well as weather forecasting models.

Verification

Even without verification, the virtual ecosystems are likely to do quite well. Verification is important to get the *details* right. The philosophy of model verification is illustrated in the slide. We start by identifying the very few features that have been reliably observed. We then:

1. Simulate those features under the conditions of the observations.
2. Compare observation and simulation.
3. Measure that matching error.
4. Adjust the model equations, often by adding new plankton guilds.
5. Iterate stages 1–5 until the matching error is minimised.

Examples of successful predictions include temperature of the mixed layer, oligotrophic zone (satellite image) and diel migration (acoustic image). Looking to the future, strong candidates include, the structure of deep chlorophyll maximum, geographical variation of rapidly adapting genes and inter-annual variability. As we seek to simulate each new set of observations made by another great experiment in biological oceanography, our model tends to become more elaborate; it includes more of the complexity of nature and more biodiversity. As usual, we follow the principle of Ockham's Razor, starting with a minimum model and only adding new equations when forced to do so by verification against good data.

Conclusion

I have demonstrated that the LE method is capable of generating virtual ecosystems that open the way to the twin goals of biological oceanography: mapping geographical and temporal variation, and explaining the distributions in terms of the basic mechanism of marine physics, chemistry and biology. I have also shown that it is possible to analyse the population dynamics of virtual ecosystems. That makes it possible to bridge the gap between marine and terrestrial ecology. There is some catching up to do, but soon I expect discoveries in virtual ecology to start contributing to theoretical ecology. Meanwhile, it provides the basis for solving the problem of under-sampling in biological oceanography.

It promises to convert biological oceanography to a branch of experimental science in which research cruises are designed to test theories by collecting data for comparison with phenomena encountered in virtual ecosystems.

Looking to the future, we have the prospect of addressing the grand challenges in biological oceanography, including:

1. Fisheries recruitment.
2. Response to ozone depletion.
3. Influence on climate change.
4. Plankton epidemiology.
5. Biodiversity.
6. And many more.

That is the challenge and fascination of virtual ecology.

Acknowledgements

I have been developing the method over the past dozen or so years with the enthusiastic help of a series of undergraduates, diploma and doctorate students, one of whom, Dr. Wolfgang Barkmann, is here today. I would like to take this opportunity to acknowledge the contribution of my team, without whom there would be no virtual ecology. I would also like to thank the many colleagues who encouraged and supported the work. Many of them are here today. Sir John Mason cannot be here because he is performing his duties as Chancellor of UMIST. When he was my doctorate supervisor, John tolerated my taking time off from cloud physics for oceanography, and later he supported the research at the Meteorological Office. I thank him and all who have been associated with the development of virtual ecology.

Appendix 1

Video on billows on internal wave.

Table 1. Slides.

	Left	Right
1	Diatoms and copepod	Fish larvae
2	Ship: HMS *Challenger*	Indian ocean plankton map
3	Satellite	Ocean colour image
4	Information content of ocean	Photo of plankton
5	Information content of observ.	Indian Ocean data
6	Photo of plankton	Map of Indian Ocean [aliased]
7	Ice age change in T & CO_2	Diatoms
8	Virtual ecosystem	Photo of plankter
9	Photo: diatoms, copepods, etc.	The model
10	Method of integration	Schematic of method
11	Research group	Automation
12	Photo of diatoms	Photo of diatoms
13	Audit trail: depth	Audit trail: ambient light
14	Audit trail: energy uptake	Audit trail: energy store, divis.
15	Demography: biomass	Demography: life expectancy
16	Demography: birth rate	Demography: death rate
17	Dye photo of internal wave	Depth of mixed layer
18	Phytoplankton biomass	Photo of diatoms
19	Irradiance	Cell division rate
20	Photo of herbivores & diatoms	Grazing rate
21	Zooplankter with eggs	Zooplankton reproduction
22	Photo: predacious copepod	Photo: *Calanus*
23	Ship: RRS *Discovery*	Acoustic: diel migration
24	Diel migration: 7 a.m.	Diel migration: 8 a.m.
25	Diel migration: 9 a.m.	Diel migration: 10 a.m.
26	Diel migration: 11 a.m	Diel migration: noon
27	Diel migration: 1 p.m.	Diel migration: 2 p.m.
28	Diel migration: 3 p.m.	Diel migration: 4 p.m.
29	Diel migration: 5 pm.	Diel migration: 6 p.m.
30	Diel migration: 7 p.m.	Diel migration: 8 p.m.

Table 1. (*Continued*)

	Left	Right
31	Map of ocean circulation	Tracks in the Sargasso Sea
32	Antilles: mixed layer depth	Antilles: temperature
33	Antilles: irradiance	Antilles: plankton growth ra
34	Antilles: nitrate concentration	Antilles: nitrate uptake rate
35	Antilles: ammonium conc.	Antilles: ammonium uptake
36	Antilles: grazing rate	Antilles: natural mortality
37	Antilles: phyto biomass	Antilles: zooplankton bioma
38	Antilles: population dynamics	Photo of copepod
39	Map of tracks in Sargasso Sea	Circuit: mixed layer depth
40	Circuit: nitrate concentration	Circuit: nitrate uptake
41	Circuit: ammonium conc.	Circuit: ammonium uptake
42	Circuit: phyto. biomass	Circuit: zooplankton biomas
43	**OVERHEAD**: circuits	–
44	Photo of diatom	Energy reserve in cell divisi
45	Natural selection: laps 1 & 2	Natural selection: laps 3 & 4
46	Plankton multiplier hypothesis	Global warming: ice age
47	Partial pressure of CO_2	Temperature
48	Mixed layer depth	Reduction in production
49	Reduction in particle flux	Impact on global warming
50	Photo of RRS Charles Darwin	Autosub crossing Atlantic
51	Model verification	Tuning model specification
52	Map of tracks in Sargasso Sea	Sea surface temperature
53	Map of sea Rover data	Photo of Sea Soar on *Poseid*
54	Seasonal migration of bloom	Poleward slope of DCM
55	Dinoflagellates (colonial)	Dinoflagellates
56	Comb jelly (red prey)	Slinky Ascidians
57	Comb jelly	Comb jelly
58	Colonial salp	Colonial salp
59	Larva of boxcrab	Dinoflagellates
60	Application	Fish larva
61	Copepod	Lobster larva with jelly fish

CHAPTER 6

PESTS , PESTICIDES AND PEST MANAGEMENT

GRAHAM A. MATTHEWS*

"What the Palmerworm left, the swarming locust has eaten. What the swarming locust left, the hopping locust has eaten, and what the hopping locust has left, the destroying locust has eaten".
Joel 1:4

Man has fought against pests since Biblical times and even today, locusts pose one of the many threats to food supplies in several countries. In developed countries, the plentiful supply of high-quality food available in supermarkets has been achieved by adopting chemical controls since the 1950s in conjunction with other improved farming practices. In contrast, in many of the developing countries, food supplies are often inadequate to provide the needs of an ever-growing human population (Weber, 1994). While lack of rainfall and other socio-economic factors are involved in reducing yields, pests (including insects, fungi, nematodes and competition by weeds), still take an unacceptable toll on crops. Some estimates indicate that more than 50% of produce grown for human consumption is lost before it reaches the consumer (Oerke *et al.*, 1994). However, since Rachel Carson's book *Silent Spring* was published in 1962, there has been world-wide concern about the effects of pesticides,

*Department of Biology, Imperial College of Science, Technology and Medicine, Silwood Park, Ascot, Berkshire, SL5 7PY.

notably the chlorinated hydrocarbon insecticides, arising from environmental pollution. This concern has led to a reappraisal of the way these chemicals should be used in conjunction with other control tactics in pest management programmes.

As an example of the development of a pest management programme, emphasis is given here to cotton, particularly in southern Africa. Cotton is a major crop grown in over 40 countries, not only for its fibre, but also as a source of vegetable oil and a protein-rich animal feed. In many countries, the crop provides employment in local processing industries and an important source of foreign exchange. It has been the target of major agrochemical companies in the development of new pesticides and, consequently, extensively treated by a wide range of different pesticides. To combat some of the problems that have arisen due to the overuse of insecticides, attention has been given more recently by these companies to the development of new cultivars toxic to insect pests by using genetic engineering in an effort to reduce the need to spray crops.

Insect Pests of Cotton

The crop provides an ideal habitat for a wide range of different insects (Pearson, 1958), but fortunately, relatively few species are major pests, and not all of these are found in every area of crop production. Following World War I, and the disruption of cotton supplies to Europe from the United States (US), growing cotton was encouraged in southern Africa in the 1920s. Seed of American varieties of *Gossypium hirsutum*, notably Cokers Wild, proved to be highly susceptible to the cotton jassids (*Jabobiella facialis*). Fortunately, the plant breeders had some cotton varieties from India, one of which, MU8A, had pubescent leaves and was resistant to the jassids. A variety, 9L34, was selected (Parnell *et al.*, 1949) and remained the mainstay of cotton production, albeit with low yields until the 1950s. The resistance to this pest by growing hairy-leaved cultivars has remained a key component of the integrated pest management (IPM) programme by enabling growers to avoid the application of insecticide sprays during the first six to eight weeks of the season before flower buds are produced (Matthews, 1989).

Once flower buds are produced, cotton plants are very vulnerable to the infestations of lepidopterous larvae, known as bollworms. Several species can cause major crop loss, but in southern Africa, the red bollworm *Diparopsis castanea* was regarded as a key pest. Legislation had been introduced in the 1930s to enforce farmers to uproot cotton stalks after harvesting and to have a two month cotton-free period before the next crop was sown. This type of closed season is now universally accepted as an essential part of IPM of cotton for pink bollworm (*Pectinophora gossypiella*) control, but as the red bollworm pupae can survive between seasons in diapause (Tunstall, 1968), the only effect of the closed season was to reduce the build-up of a pest population on ratoon or standover plants before the onset of rain when the new season's crop is sown. Some initial attempts at using chemical control had shown that yields could be significantly increased, but if insecticides were to be used as effectively and economically as possible, the first requirement was to understand as much as possible about the biology of the pest to identify the most vulnerable stage in its life cycle.

Detailed observations from bollworm egg hatch to moth emergence under field conditions revealed that the only time the larvae were really exposed on the plant surface was the initial period after hatching until the first instar larva bored into a flower bud or boll (Tunstall, 1961). This period could be as short as a few minutes, particularly if an egg was laid on one of the bracts surrounding a bud, to two to three hours. Hairs at the junction of the leaf lamina and the petiole confused some larvae which failed to escape from a leaf. Once the larva had penetrated a bud or boll, it seldom, if ever, reappeared until much more fully grown. Consequently, the only time at which the larvae would be exposed to an insecticide deposit was during the first instar. Laboratory and subsequent single plant assessments of different insecticides were with the first instar larvae from egg hatching (Matthews, 1966).

Prior to these studies, the devastating effect of infestations of another bollworm, the so-called "American bollworm" (*Helicoverpa armigera*), named after the species in Mississippi in the US, had been noted in some seasons. It was soon apparent that if the red bollworm was controlled, the improvement in survival of buds and increase in flowering attracted more *H. armigera* infestations. The evaluation of insecticides had, therefore, to be extended and soon major differences in the susceptibility of the two bollworms were measured. At that time, in the early 1960s, dichloro diphenyl trichloroethane (DDT) was considered a safe insecticide to use, having been used to treat 2.5 million humans during World War II to control the louse-borne typhus outbreak in Naples and prevent its spread to troops (Brown, 1951). It was, therefore, selected as the most effective against *H. armigera* whereas the carbamate carbaryl was significantly more effective than DDT against *D. castanea*. The agrochemical industry responded to this information very rapidly by marketing a mixture of these two insecticides. However, a detailed analysis of the records of infestations during several seasons indicated that both bollworms seldom, if ever, required treatment at the same time. To use a mixture of insecticides would be uneconomic, especially as the crop is grown by a large number of small-scale farmers. Normally, the early red bollworm infestation declined during peak flowering when *H. armigera* was the main pest. Later, at the end of the season, red bollworm could again be more serious. This pattern of infestation led to the concept of dividing the season into three sections, using carbaryl from about six weeks after germination, then DDT and later reverting back to carbaryl, which also proved to be very effective against the cotton stainers, *Dysdercus* spp. The problem was when to make the changes, especially as the onset of the *H. armigera* infestation was shown to vary from mid-January to early March, probably in relation to when a generation emerges from maize and indirectly to rainfall. Field trials with applications to coincide with different populations of bollworm eggs showed that yields were significantly reduced if *H. armigera* egg counts exceeded 0.5 per plant (Matthews & Tunstall, 1968).

Earlier, field trials located on farmers' fields in the Shire Valley and Lake Shore areas of Nyasaland (now Malawi) and Rhodesia (now Zimbabwe) involved twice weekly bollworm egg counts to time spray

Table 1. Increase in yields of seed cotton in Central Africa.

	Year	Yield Seed Cotton (kg/ha)		Percentage Stained and Damaged	
		Treated	Untreated	Treated	Untreated
Malawi	1960	1429	560	7.6	19.6
	1961	1323	574	8.1	28.7
Zimbabwe	1960	2172	1361	5.9	45.6
	1961	2174	579	0.9	11.6

applications, instead of recommending a weekly schedule regardless of pest populations. The trials were very successful and demonstrated significant yield increases due to spraying (Table 1) (Tunstall & Matthews, 1966). The best increases on individual farms were generally related to soil fertility, rainfall and weed management. The recommendations following these trials are still valid, except that DDT has been replaced by the synthetic pyrethroids, notably fenvalerate, bifenthrin and lambda-cyhalothrin. Certain pyrethroids have not been recommended as, like DDT, these tended to encourage an increase in red spider mite populations. The restriction of pyrethroid applications to a relatively short period of a maximum of nine weeks each year, as soon as they were introduced in 1979/80 (Brettell, 1986), was considered to be one of the factors reducing the selection for bollworms with resistance to this group of insecticides. This contrasts with the situation in Australia, where pyrethroid resistance to *H. armigera* was apparent before a three-stage programme was introduced as a resistance management strategy (Forrester *et al.*, 1993). The cautious approach to pyrethroids in Zimbabwe was a reflection of the problems encountered earlier when red spider mites became resistant to the organophosphate dimethoate, then used as an acaricide. This problem had been overcome by dividing Zimbabwe into distinct cotton areas, within which the use of a limited number of acaricides would be rotated to minimise the use of a chemical with a particular mode of action (Duncombe, 1973).

Fig. 1. Scouting for bollworm eggs using a pegboard.

The insect control recommendations (Tunstall & Matthews, 1961) have stood the test of time and Zimbabwe's annual production of seed cotton has increased to about 200,000 tonnes (Matthews, 1989).

Many considered that a complex set of recommendations could not be used by small-scale farmers. Initially, selected extension staff employed by the government were trained to do the scouting for bollworm eggs on a sample of farmers' fields. This data was provided in the nearest village so that individual farmers in the area could follow the recommendations. General data was also provided in a short weekly radio programme. This was followed by the development of a simple pegboard (Beeden, 1972) based on a sequential sampling system so that farmers could do their own scouting. Subsequently, a school was established in Zimbabwe to train farmers in insect identification and how to decide what to spray (Burgess, 1983).

Application Technology

The adoption of chemical control by small-scale farmers with little or no capital investment in Malawi necessitated that equipment for insecticide application should be as inexpensive as possible. Manually operated knapsack sprayers had been designed a century earlier to treat vines with Bordeaux mixture to control downy mildew disease (Lodeman, 1896), but could African farmers apply these new insecticides safely? Long before control of substances hazardous to health (COSHH) assessments were required in the United Kingdom (UK), there was a clear need to avoid the most toxic insecticides, especially as the full protective clothing recommended by the agrochemical companies was too expensive and totally unsuitable for farmers walking in the heat of the Shire Valley. Using a knapsack sprayer has traditionally involved waving a nozzle on the end of lance in front of the operator walking between the crop rows being treated. The body inevitably brushed into the treated foliage. Hence operator contamination would be unacceptable. To overcome this problem and also distribute the insecticide more evenly, though the crop canopy where the bollworm larvae were walking towards buds or bolls, a rear-mounted "tailboom" was developed (Tunstall *et al.*, 1961). The number of nozzles and the volume of spray was increased in relation to plant height during the season. Hence with the concentration of the spray kept constant, the dosage was approximately proportional to the leaf area. The more efficient distribution of the insecticide directed specifically at the first instar larvae allowed a lower dosage to be applied in comparison with the rates recommended in the US and elsewhere (Matthews, 1994).

Fig. 2. Spraying cotton with knapsack fitted with a "tailboom".

Rotary atomisers

Promotion of the tailboom on knapsack sprayers in Malawi (Gower & Matthews, 1971) was only partially successful as farmers had great difficulty in obtaining sufficient water to their fields. This led to research on minimising the spray volume. Sayer (1959) had developed an ultra-low volume method of applying insecticide to treat locust hopper bands. Subsequently, Edward Bals introduced a light-weight hand-carried battery-operated spinning disc sprayer, and this unit was used in field trials on cotton in Malawi (Matthews, 1973). Initial work was with oil-based formulations of the same insecticides used in the knapsack sprayer, but applied at up to 2.5 litres per hectare, the volume increasing as the number of rows treated by each pass across a field was decreased from six to four and then two rows to compensate for the increase in leaf area. The technique involved releasing a spray containing more uniform-sized droplets above the crop canopy and allowing the air movement

Fig. 3. ULV spraying with a hand-carried spinning-disc sprayer in Africa.

across the crop and turbulence created to disperse the spray within the crop canopy. The involatile formulation prevented the droplets from shrinking in size. Thus, although there was a downwind movement of the spray, droplets were deposited within a few metres from the sprayer. Yields comparable with knapsack spraying were obtained and subsequently, the technique of ultra-low volume (ULV) spraying of cotton was adopted most extensively in francophone West Africa because farmers refused to use knapsack sprayers (Cauquil, 1987).

Later, studies with water-based formulations and with mixtures containing molasses showed that the use of very low volumes (five to 15l/ha) (Matthews, 1990) was equally effective and allowed greater versatility in the choice of insecticide. The use of the less expensive water-based formulations has enabled the very-low volume (VLV) spraying technique to supersede ULV applications, particularly as the hand-held equipment has been developed with a smaller disc and improved motor so that battery consumption has been reduced significantly (Clayton, 1992). Using the same principle, a disc rotated at a slower speed to produce large droplets, not prone to downwind drift, has been used in the controlled droplet application (CDA) of herbicides, especially in amenity/urban areas and in some overseas plantations. CDA technology, however, has not had an impact on UK agriculture, where the availability of water is not a problem and the atomiser costs more than the traditional pressure nozzles.

Electrostatic sprays

There was concern that using rotary atomisers, deposition within the crop canopy was dependent on air turbulence to impact droplets on the foliage and other parts of the plants. To determine whether an additional force could improve deposition with less risk of downwind drift, investigations on electrostatic spraying were initiated. Work in Sheffield by Hopkinson (1974) was followed up by studies at Rothamsted using rotary atomisers (Arnold & Pye, 1980; Cayley *et al.*, 1984), hydraulic nozzles (Marchant & Green, 1982; Marchant *et al.*, 1985) and rotary atomisers (Marchant, 1985). At the same time, concerned mainly by the requirements of small-scale farmers in semi-arid areas, a truly ultra-low volume electrostatic sprayer, the "Electrodyn", was developed by Coffee (1979) to apply less than one litre of oil-based formulation per hectare. Subsequently, it was used commercially in Brazil, Angola, Mozambique, Nigeria and Tanzania to treat cotton and, to a lesser extent, cowpea crops (Smith, 1988, 1989; Matthews, 1989, 1990). The insecticide was safely packaged in the factory in the "Bozzle", a plastic container

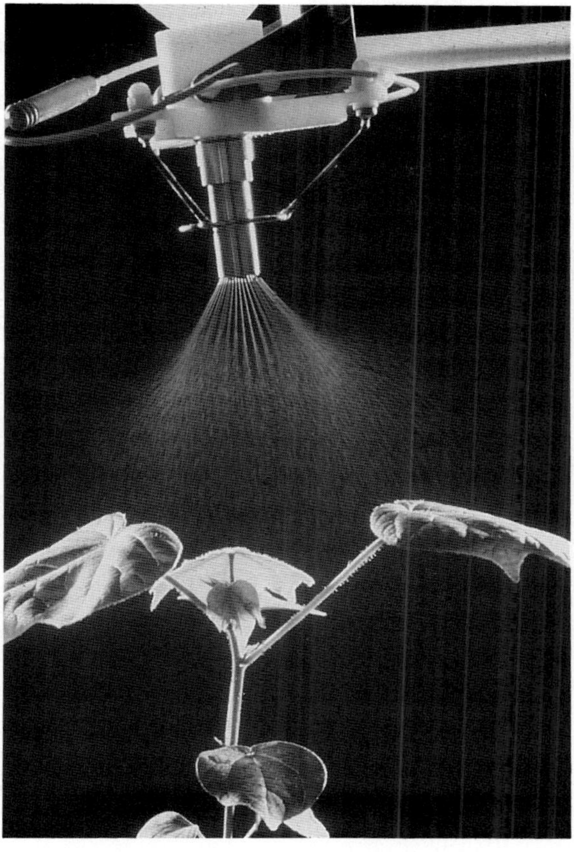

Fig. 4. Prototype "Electrodyn" sprayer for ULV spraying of cotton.

which incorporated the electrodynamic nozzle. This eliminated any pouring out or measuring quantities of concentrated pesticide by the operator.

Unfortunately, the technique has not been successful commercially as farmers do not want to purchase their pesticides from one multi-national agrochemical company, nor do they want to have a restricted list of chemicals which can be formulated in such a low volume. Nevertheless, the potential for an electrostatic spray in an IPM programme

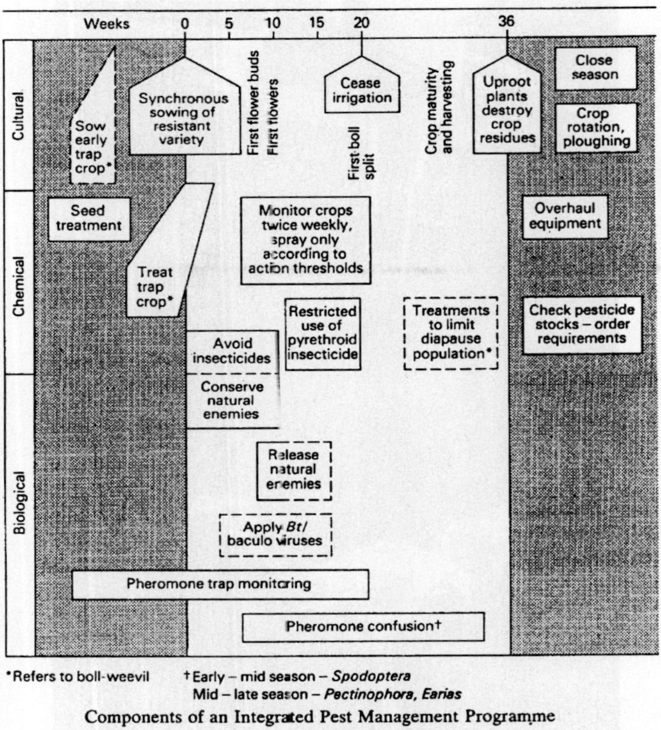

Fig. 5. Generalised IPM programme for cotton, based on experience in Central Africa.

remains as it is possible to improve deposition and reduce the amount of active ingredient applied in the crop. When used on strips of crops as a modification of traditional inter-cropping, the use of electrostatic sprays could enhance the survival of natural enemies in the soil and on adjacent untreated crops.

Knapsack sprayer development

While research on reduced volumes of spray continued, attention was also directed at the design of lever-operated knapsack sprayers to ensure

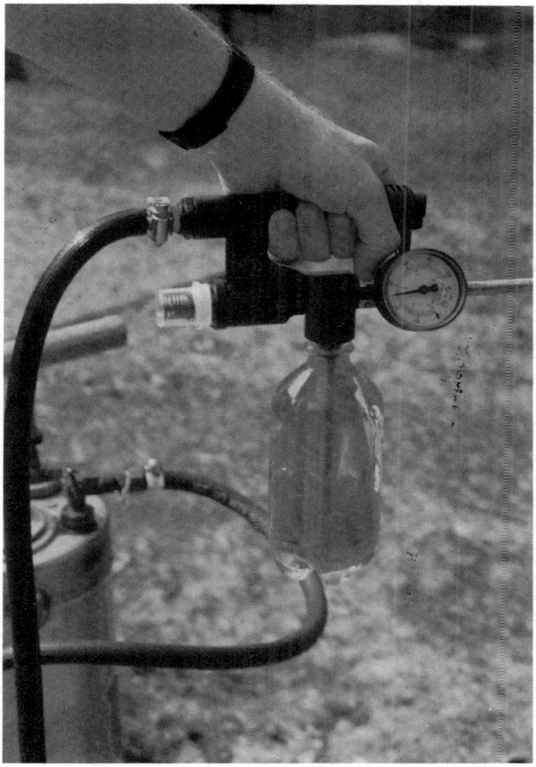

Fig. 6. Manually operated crop sprayer using Down Fluid Injection method — DFI.

that they were more durable (Matthews *et al.*, 1969) and safer to use. The legislation in the UK, particularly the Health and Safety at Work and the Food and Environment Protection Act, provided the stimulus for standards of equipment and working practices that manufacturers and farmers had to meet. The preparation of sprays has long been recognised as the time at which most spray operators are most exposed to contamination by pesticides. In Malawi, the small sachets of insecticide were developed to contain sufficient formulation for each knapsack load. (Gower & Matthews, 1971) and attempts were made to use water-soluble sachets, a technique now widely used by the agrochemical industry. In the UK, tractor-powered equipment, the provision of engineering controls, such as low-level mixing units and closed transfer systems, to protect the user has taken priority, but relatively little attention has been given to the same problem for small-scale farmers in the tropics. The packaging of concentrated liquid formulations in containers with a built-in measure has been introduced on a limited scale (Pflazer, 1993). A closed container, with a bag inside to contain the pesticide, has been designed by Laurie Down in Australia. The disposable container dispenser was designed to fit manually operated sprayers and inject pesticide into the flow of water to the nozzle (Craig *et al.*, 1993), but has not yet been commercially accepted. As the pressure at the nozzle on manually operated sprayers varies, part of the system is a valve — referred to as a Spray Management Valve (SMV) — to provide a fixed, controlled pressure. This was considered to be especially important when applying herbicides to avoid sprays being applied at too high a pressure and droplets drifting onto susceptible crops downwind (Matthews & Thornhill, 1993).

Choice of nozzle

Farmers had often used only one type of nozzle as supplied when the equipment was purchased. Following the concern of spray drift from cereal fields affecting horticultural crops (Elliott & Wilson, 1983), the British Crop Protection Council's (BCPC) Applications Committee initiated, with International Pesticide Application Research Centre (IPARC) in Imperial College, analyses of a selected range of nozzles by laboratories in Europe and the US, using different measuring systems (Arnold, 1987). The data was subsequently used to develop a spray classification system (Fig. 7) (Doble *et al.*, 1985). As there are differences in the numerical data between measuring systems, reference nozzles were selected to indicate the droplet spectra representative of each category of spray. This spray classification forms part of the Codes of Practice in the UK, and has now been adopted in Europe and the US. Equipment manufacturers can provide farmers with the appropriate information on the quality of spray

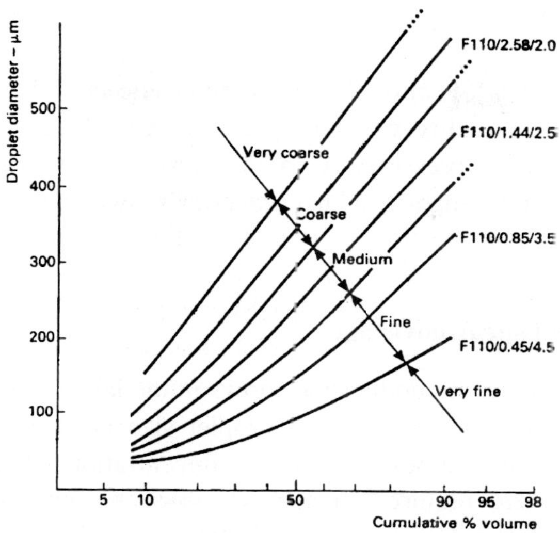

Fig. 7. Spray classification scheme adopted in the UK.

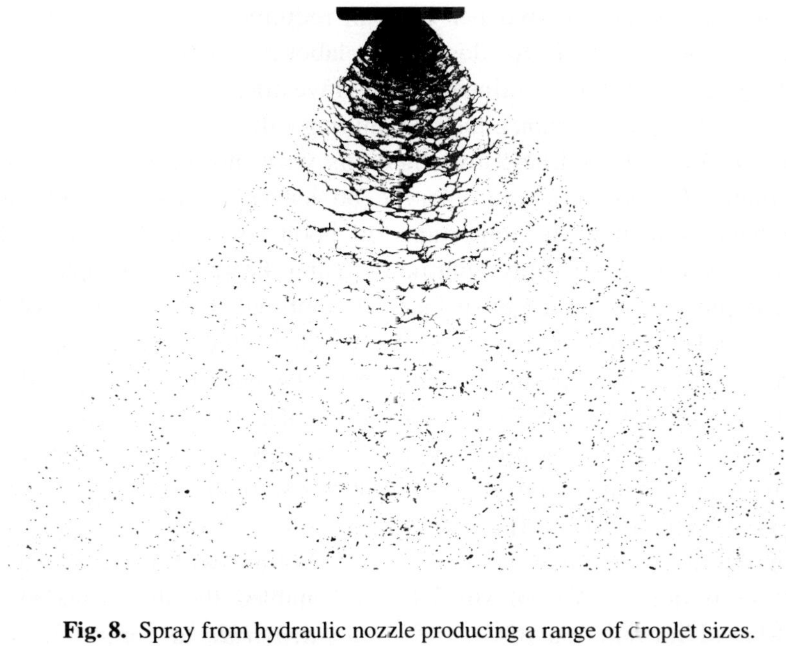

Fig. 8. Spray from hydraulic nozzle producing a range of droplet sizes.

from their nozzles, while agrochemical companies can either recommend a particular spray quality or more precisely recommend a particular nozzle and operating pressure. A BCPC code for identifying nozzles, without specific reference to a manufacturer's own identification, was also introduced.

Droplet size and spray coverage

The ability to produce monodisperse sprays in the laboratory has enabled more basic studies on the optimum droplet size needed on foliage to control sessile or mobile pests, the spray concentration needed, and how many droplets were required to achieve adequate spray coverage. In collaboration with Nigel Scopes at the Glasshouse Crops Research Station, Munthali (1984) investigated the effect of the acaricide dicofol on the red spider mite *Tetranychus urticae* eggs (Munthali & Scopes, 1982). He showed that the smallest droplets used (20 μm diameter), at a concentration of one to two per cent ai, required the least amount of pesticide to achieve an LD_{50} dose in the laboratory (Munthali & Wyatt, 1986). Abdalla (1984) continued these investigations with whiteflies (*Trialeurodes vaporarorium*), also examining differences in formulation (Adams *et al.*, 1987). Later investigations with more mobile diamond back moth (*Plutella xylostella*) larvae on leaves (Omar & Matthews, 1987) have confirmed the importance of good coverage, which can be achieved with the least spray volume by using the smallest droplets that can be deposited on the target surface. The optimal small droplets would be extremely prone to drift, and hence the interest in air-assisted electrostatically charged sprays that were subsequently investigated by Abdelbagi and Adams (1987). Significantly much lower volumes of spray can be applied to most crops compared to the high volume treatments used in many countries on horticultural crops, such as tomatoes, with far less wasted as "run-off" to the soil.

Optimum droplet size is particularly important with the application of mycopesticides. Recent studies have enabled the development of an oil-based formulation of *Metarhizium flavoviride* to control locusts

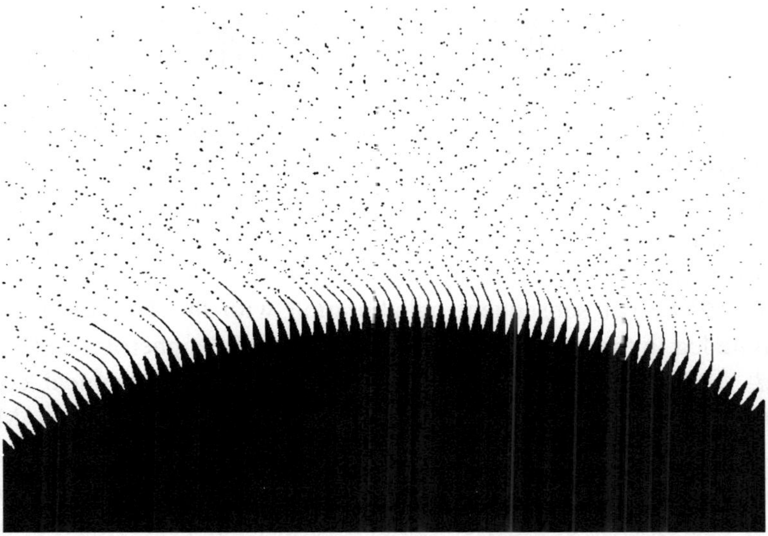

Fig. 9. Spray from spinning disc.

(Bateman, 1994; Bateman *et al.*, 1994). Too small a droplet delivers too few spores to the locust, whereas valuable formulation is wasted if droplets are too large. Hence, ideally, the mycoinsecticide is applied in sprays with droplets of 70 to 90 μm, for example with a spinning disc (Fig. 9).

Integrated Pest Management

Four decades ago, many entomologists felt that the arrival of organic insecticides would solve all the pest problems, but as Ripper (1944) pointed out right at the beginning, pests can soon become resistant to the pesticides and, in the absence of natural enemies, pest populations can resurge after chemical treatments. Hence biological and other control tactics need to be integrated in as harmonious a manner as possible. Even with careful development of a spray programme on cotton in Central Africa, as pointed out above, mites became resistant to dimethoate, probably because the same chemical was used throughout the year on a succession of crops grown under irrigation on some farms.

In addition to the use of jassid-resistant cotton cultivars in Africa, the possibilities of using a number of control tactics were investigated. Early observations of catches of up to 1,000 *Diparopsis castanea* male moths attracted to a trap baited with seven virgin female moths were not then considered likely to lead to a method of control (Tunstall *et al.*, 1959), but subsequent collaboration with the Tropical Products Institute led to the identification and synthesis of the pheromone (Nesbitt *et al.*, 1975). Subsequently, its use in traps to time chemical control treatments (Marks, 1977) and the possibility of mating disruption (Marks *et al.*, 1977, 1981) were investigated. Although this work has not resulted in large-scale adoption, these studies initiated interest in the use of pheromones of other cotton bollworms. The deployment of hand-applied pheromone dispensers to confuse pink bollworm (*Pectinophora gossypiella*) moths in Egypt on an extensive area-wide scale has allowed a reduction in insecticide sprays (Gillham *et al.*, 1995). Other "lure and kill" techniques have been tried in the US to control the boll weevil (Smith *et al.*, 1995).

In Zimbabwe, the impact of certain predators, notably the ant *Pheidole*, was recognised, but little work was done on biological control, following attempts back in the 1930s to use trap crops and release *Trichogramma*. In contrast, the Russian entomologists' reaction to the devastating pollution of the Aral Sea area by the aerial application of organophosphate and other insecticides in the 1960s and 1970s initiated the rearing of *Trichogramma pintoi* on *Sitotroga cerealella* eggs (Filippov, 1992). This technique has been used in over 700 bio-factories located throughout the cotton areas of Uzbekistan. There is much interest in other countries, but much depends on the economic production of the parasitoids and the intensity of pest infestations. In Central Asia, at least the severe winter has a major impact on pest survival and as there is no longer widespread use of chemical pesticides, the natural enemy populations are effective.

The public perception is that all pesticides are highly poisonous to man, but since the 1950s, the requirements of registration authorities, such as the Environmental Protection Agency in the US, and the search for chemicals with different modes of action has led to the marketing of many less toxic pesticide products. The acylyurea insect growth regulators are more specific to insects and less hazardous to man. Re-registration requirements are forcing some of the agrochemical companies to withdraw some of the older products, at least in the developed countries. Unfortunately, in the promotion of free trade, many generic pesticides which are no longer covered by patents are still readily available in developing countries and marketed much like soap powders with little advice on their correct application in the field. There is clearly a need to regulate the use of the most hazardous pesticides (Gillham *et al.*, 1995). Perhaps, like the pharmaceutical industry with "prescription only medicines", there should be "prescription only pesticides" available only to licensed/qualified applicators. In the UK, users of pesticides on farms now have to pass a practical test, while other countries are introducing regulatory checks on the equipment used by farmers. In some areas, the limited use of pesticides has been possible within specified areas, such as the Zimbabwean acaricide rotation scheme. These approaches will all help to avoid an overuse of pesticides, reduce the selection of resistant pest populations and minimise health problems.

Much attention has been focused on the biotechnology revolution and the hope that by genetic engineering, the addition of the gene for *Bacillus thuringiensis* toxin into crop plants will make them poisonous to insect pests. The lepidopterous larvae susceptible to *Bt* toxin may be effectively controlled for a period by the genetically engineered plants, but as with pesticides, there is an immediate risk that the pests will soon become resistant to the new cultivars. The risk is so great that scientists have begun discussing resistance management strategies before the new cultivars are available commercially (Galloway, 1994).

The future for controlling our pests undoubtedly lies in the use of IPM. Some of the old recipes of the last century (Lodeman, 1896) are being revived. In particular, the use of soap sprays against aphids and mineral oil emulsions against whiteflies are effective and less damaging to the natural enemies. Their use, in particular, on crops like melons will decrease the selection of resistance by populations spreading to other crops.

Although simple solutions are preferred, the single approach, whether it is the use of pesticides, the growing of resistant cultivars or the release of natural enemies, has proven unworkable. In the complex ecological situation in the farmers' fields, with a range of crops, pests, natural enemies, uncertain growing conditions and environmental changes, it is essential to weave together several tactics — cultural, biological and chemical — to control pests, even if each tactic is based on complex technologies. The principles outlined above, with particular reference to cotton crops and application technology, apply more widely to other crops and techniques. The problem is to deliver the package of tactics in as simplified a form as possible or as a sequence of additional tactics that can be understood and adopted by farmers. Greater adoption of pest management will require close collaboration of many farmers within an agroecological area so that each benefits from the scale of the adoption of the technique. This was seen many decades ago with cultural controls, but is equally true with new technologies, such as the release of natural enemies, pheromone treatments and resistance management strategies. Even if scientists do develop appropriate packages, there are severe constraints on their adoption due to finance and the training that is required. However, sustained collaborative and dynamic efforts to implement effective and economic pest management will have considerable benefits to environmental safety and human health.

References

Abdalla, M. (1984) *A Biological Study of the Spread of Pesticides in Small Droplets.* PhD thesis. University of London.

Abdelbagi, H.A. & Adams, A.J. (1987) Influence of droplet size, air-assistance and electrostatic charge upon the distribution of ultra-low volume sprays on tomatoes. *Crop Protection*, **6**, 226–233.

Adams, A.J., Abdalla, M.R., Wyatt, I.J. & Palmer, A. (1987) The relative influence of the factors which determine the spray droplets density required to control glasshouse whitefly, *Trialeurodes vaporarorium. Aspects of Appl. Biol.*, **14**, 257–266.

Arnold, A.C. (1987) The drop size of the spray from agricultural fan spray atomisers as determined by a Malvern and the Particle Measuring System (PMS) instrument. *Atomisation and Spray Technol.*, **3**, 155–167.

Arnold, A.J. & Pye, B.J. (1980) Spray application with charged rotary atomisers. *BCPC Monograph*, **24**, 109–125.

Bateman, R. (1994) Performance of myco-insecticides: importance of formulation and controlled droplet application. *BCPC Monograph*, **59**, 275–284.

Bateman, R., Price, R.E., Muller, E.J. & Brown, H.D. (1994) Controlling brown locust hopper bands in South Africa with a myco-insecticide spray. In *Proc. Brighton Crop Protection Conf.: Pests and Diseases*, pp. 609–616.

Beeden, P. (1972) The pegboard — an aid to cotton pest scouting. *PANS*, **18**, 43–45.

Brettell, J.H. (1986) Some aspects of cotton pest management in Zimbabwe. *Zimbabwe Agric. J.*, **83**, 41–46.

Brown, A.W.A. (1951) *Insect Control by Chemicals.* New York: Wiley.

Burgess, M.W. (1983) Development of cotton pest management in Zimbabwe. *Crop Protection*, **2**, 247–250.

Cauquil, J. (1987) Cotton pest control: a review of the introduction of ultra-low volume (ULV) spraying in sub-Saharan French-speaking Africa. *Crop Protection*, **6**, 38–42.

Cayley, G.R., Etheridge, P., Griffiths, D.C., Phillips, F.T., Pye, B.J. & Scott, G.C. (1984) A review of the performance of electrostatically charged rotary atomisers on different crops. *Ann. appl. Biol.*, **105**, 379–386.

Clayton, J. (1992) New developments in controlled droplet application (CDA) techniques for small farmers in developing countries — opportunities for formulation and packaging. In *Proc. Brighton Crop Protection Conf.*, **1**, 333–342.

Coffee, R.A. (1979) Electrodynamic energy — a new approach to pesticide application. In *Proc. Brighton Crop Protection Conf. — Pests and Diseases*, **3**, 777–789.

Craig, I.P., Matthews, G.A. & Thornhill, E.W. (1993) Fluid injection metering system for closed pesticide delivery in manually operated sprayers. *Crop Protection*, **12**, 549–553.

Doble, S.J., Matthews, G.A., Rutherford, I. & Southcombe, E.S.E. (1985) A system for classifying hydraulic nozzles and other atomisers into categories of spray quality. In *Proc. Brighton Crop Protection Conf. — Weeds*, 1125–1133.

Duncombe, W.C. (1973) The acaricide spray rotation for cotton. *Rhod. Agric. J.*, **70**, 115–118.

Elliott, J.G. & Wilson, B.J. (1983) The influence of weather on the efficiency and safety of pesticide application: the drift of herbicides. *BCPC Occasional Bulletin*, **3**.

Filippov, N.A. (1992) Place of bio-control in integrated pest management in the former Union of Soviet Socialists Republics. In *Cotton-Integrated Pest Management: Proc. Symp.* (September 1990 USDA ARS-106), pp. 1–16.

Forrester, N.W., Cahill, M., Bird, L.J. & Layland, J.K. (1993) Management of pyrethroid and endosulfan resistance in *Helicoverpa armigera* (Lepidoptera: Noctuidae) in Australia. *Bull. ent. Res.*, Supplement No. 1 IIE, London.

Gannaway, J.R. (1994) Breeding for insect resistance. In *Insect Pests of Cotton*, eds. G.A. Matthews & J.P. Tunstall, pp. 431–453. Wallingford: CABI.

Gillham, F.E.M, Bell, T.M., Arin, T., Matthews, G.A., Le Rumeur, C. & Hearn, A.B. (1995) *Cotton Production Prospects for the Next Decade.* World Bank Technical Paper No. 287, Washington DC.

Gower, J. & Matthews, G.A. (1971) Cotton development in the southern region of Malawi. *Cott. Gr. Rev.*, **48**, 2–18.

Hopkinson, P.R. (1974) The prospects of enhanced impaction of fine sprays by electrostatic charging. *Br. Crop Protection Monograph*, **11**, 166–179.

Lodeman, E.G. (1896) *The Spraying of Plants.* London: Macmillan.

Marchant, J.A. (1985) An electrostatic spinning disc atomiser. *Trans. Am. Soc. Ag. Engrs.*, **28**, 386–392.

Marchant, J.A. & Green, R. (1982) An electrostatic charging system for hydraulic spray nozzles. *J. Agric. Engng. Res.*, **27**, 309–319.

Marchant, J.A., Dix, A.J. & Wilson, J.M. (1985) The electrostatic charging of spray produced by hydraulic nozzles — Part II: Measurements. *J. Agric. Engng. Res.*, **31**, 345–360.

Marks, R.J. (1977) Assessment of the use of sex pheromone traps to time chemical control of the red bollworm *Diparopsis castanea* Hampson (Lepidoptera: Noctuidae) in Malawi. *Bull. ent. Res.*, **67**, 575–587.

_________ (1978) The influence of pheromone trap design and placement on the catch of the red bollworm *Diparopsis castanea* Hampson (Lepidoptera: Noctuidae). *Bull. ent. Res.*, **68**, 31–45.

Marks, R.J., Nesbitt, B.F., Hall, D.R. & Lester, R. (1978) Mating disruption of the red bollworm of cotton *Diparopsis castanea* Hampson (Lepidoptera: Noctuidae) by ultra-low volume spraying with a microencapsulated inhibitor of mating. *Bull. ent. Res.*, **68**, 11–29.

Matthews, G.A. (1966) Investigations of the chemical control of insect pests of cotton in Central Africa. I: laboratory rearing methods and tests of insecticides by application to bollworm eggs. II: tests of insecticides with larvae and adults. *Bull. ent Res.*, **57**, 69–91.

_________ (1973) Ultra-low volume spraying of cotton in Malawi. *Cott. Gr. Rev.*, **50**, 242–267.

_________ (1989) *Cotton Insect Pests and Their Management.* Harlow: Longman.

_________ (1990) Changes in application techniques used by the small-scale cotton farmer in Africa. *Tropical Pest Management*, **36**, 166–172.

_________ (1994) Chemical control. In *Insect Pests of Cotton*, eds. G.A. Matthews & J.P. Tunstall, pp. 535–557. Wallingford: CABI.

Matthews, G.A., Higgins, A.E.H. & Thornhill, E.W. (1969) Suggested techniques for assessing the durability of lever-operated knapsack sprayers. *Cott. Gr. Rev.*, **46**, 143–148.

Matthews, G.A. & Thornhill, E.W. (1993) Herbicide application: equipment design for small-scale farmers. *Brighton Crop Protection Conf. — Weeds*, pp. 1171–1176.

Matthews, G.A. & Tunstall, J.P. (1968) Scouting for pests and the timing of spray application. *Cott. Gr. Rev.*, **45**, 115–127.

Munthali, D.C. (1984) Biological efficiency of small dicofol droplets against *Tetranychus urticae* (Koch) eggs, larvae and protonymphs. *Crop Protection* **3**, 327–334.

Munthali, D.C. & Scopes, N.E.A. (1982) A technique for studying the biological efficiency of small droplets of pesticide solutions and a consideration of the implications. *Pesticide Sci.* **13**, 60–62.

Nesbitt, B.F., Beevor, P.S., Cole, R.A., Lester, R. & Poppi, R.G. (1975) The isolation and identification of the female sex pheromones of the red bollworm moth *Diparopsis castanea. J. Insect Physiol.*, **21**, 1091–1096.

Oerke, E.-C., Dehne, H.-W., Schonbeck & Weber, A. (ed., 1994) *Crop Production and Crop Protection.* Amsterdam: Elsevier.

Omar, D. & Matthews, G.A. (1987) Biological efficiency of spray droplets of permethrin ULV against diamond back moth. *Aspects of Appl. Biol.*, **14**, 173–179.

Parnell, F.R., King, H.E. & Ruston, D.F. (1949) Insect resistance and hairiness of the cotton plant. *Bull. ent. Res.*, **39**, 539–575.

Pearson, E.O. (1958) *The Insect Pests of Cotton in Tropical Africa.* Emp. Cott. Gr. Corp., London and Commmonwealth Inst. Ent., London.

Pflazer, H. (1993) Safety aspects and legislation trends. In *Application Technology for Crop Protection*, ed. G.A. Matthews & E.C. Hislop, pp. 13–33. Wallingford: CABI.

Ripper, W.E. (1944) Biological control as a supplement to chemical control of insect pests. *Nature*, **153**, 448–452.

Sayer, H.J. (1959) An ultra-low volume spraying technique for the control of desert locust *Schistocerca gregaria. Bull. ent. Res.*, **50**, 371–386.

Smith, J.W., McKibben, G.H., Villavaso, E.J., McGovern, W.L. & Jones, R.G. (1995) Management of the cotton boll weevil with attract-and-kill devices. In *Challenging the Future: Proceedings of the World Cotton Conference 1*, ed. G.A. Constable & N.W. Forrester, pp. 480–484. Brisbane, Australia.

Smith, R. (1988) The "Electrodyn" sprayer: matching the technology to contrasting areas of smallholder agriculture. *Chemistry and Industry*, **6**, 196–199.

_________ (1989) The "Electrodyn" sprayer as a tool for rational pesticide management in smallholder cotton In *Pest Management in Cotton*, ed. M.B. Green & D.J. de B. Lyon, pp. 227–246. Chichester: Ellis Horwood.

Tunstall, J.P. (1961) The behaviour and natural mortality of Diparopsis under field conditions. In *Annual Report Cotton Pest Research Scheme for 1959/60, Part II,* pp. 69–70. Malawi Ministry of Agriculture.

_________ (1968) Pupal development and moth emergence of the red bollworm (*Diparopsis castanea* Hmps.) in Malawi and Rhodesia. *Bull. ent. Res.*, **58**, 289–299.

Tunstall, P. & Matthews, G.A. (1961) Cotton insect control recommendations for 1961–62 in the Federation of Rhodesia and Nyasaland. *Rhod. Agric. J.*, **58**, 289–299.

———— (1966) Large-scale spraying trials for the control of cotton insect pests in Central Africa. *Cott. Gr. Rev.*, **43**, 121–139.

Tunstall, J.P., Sweeney, R.C.H. & Matthews, G.A. (1959) Cotton insect pest investigations in the Federation of Rhodesia and Nyasaland. I: cotton bollworm investigations. *Cott. Gr. Rev.*, **36**, 268–275.

Weber, A. (1994) Population growth, agricultural production and food supplies. In *Crop Production and Crop Protection*, eds. E.-C. Oerke, H.-W. Dehne & A. Weber, pp. 1–44. Amsterdam: Elsevier.